NATIONAL FORENSIC SCIENCE AND
TIGATION SERIES

: Max Houck

e Law and Forensic Ballistics

48404322

amination of Documents:
d techniques, 2nd edition

48405800

vestigation of Explosions

7484 05657

amination of Human Hair

748405671

xamination of Fibres,

n and M Grieve
748408160

xamination of Glass and Paint:
nd interpretation

748405794

Speaker Identification

415 27182 7

Evidence
n
824754143

The Practice of Crime Scene Investigation
J Horswell
ISBN 9780748406098
2004

Fire Investigation
N Nic Daéid
ISBN 9780415248914
2004

Fingerprints and Other Ridge Skin Impressions
C Champod, C J Lennard, P Margot, and M Stoilovic
ISBN 9780415271752
2004

Firearms, the Law, and Forensic Ballistics, Second Edition
Tom Warlow
ISBN 9780415316019
2004

Forensic Computer Crime Investigation
Thomas A. Johnson
ISBN 9780824724351
2005

Analytical and Practical Aspects of Drug Testing in Hair
Pascal Kintz
ISBN 9780849364501
2006

Nonhuman DNA Typing: Theory and Casework Applications
Heather M Coyle
ISBN 9780824725938
2007

INTER
INVES
Series Edito

Firearms, th
T A Warlow
ISBN 97807
1996

Scientific E
methods ar
D Ellen
ISBN 97807
1997

Nonhuman I

Theory and
Applica

Forensic In
A Beveridg
ISBN 97807
1998

Forensic E
J Robertso
ISBN 97807
1999

Forensic E
2nd editio
J Robertso
ISBN 9780
1999

Forensic E
analysis a
B Caddy
ISBN 9780
2001

Forensic S
P Rose
ISBN 978
2002

Bitemark
B. J. Dori
ISBN 978
2004

INTERNATIONAL FORENSIC SCIENCE AND INVESTIGATION SERIES

Nonhuman DNA Typing

Theory and Casework Applications

Edited by
Heather Miller Coyle

CRC Press
Taylor & Francis Group
Boca Raton London New York

CRC Press is an imprint of the
Taylor & Francis Group, an informa business

CRC Press
Taylor & Francis Group
6000 Broken Sound Parkway NW, Suite 300
Boca Raton, FL 33487-2742

Printed in the United States of America on acid-free paper
10 9 8 7 6 5 4 3 2 1

International Standard Book Number-10: 0-8247-2593-X (Hardcover)
International Standard Book Number-13: 978-0-8247-2593-8 (Hardcover)

Library of Congress Cataloging-in-Publication Data

Nonhuman DNA typing : theory and casework applications / editor, Heather M. Coyle.
p. cm. -- (International forensic science and investigation series)
Includes bibliographical references and index
ISBN 978-0-8247-2593-8 (alk. paper)
1. Forensic biology. 2. DNA fingerprinting. 3. Animal genetics. I. Coyle, Heather Miller. II. Title. III. Series.

QH313.5.F67N66 2007
614'.1--dc22 2007008081

Visit the Taylor & Francis Web site at
http://www.taylorandfrancis.com

and the CRC Press Web site at
http://www.crcpress.com

Dedication

For my father, George M. Miller, who always encouraged me to succeed...

Table of Contents

International Forensic Science Series

The modern forensic world is shrinking. Forensic colleagues are no longer just within a laboratory but across the world. E-mails come in from London, Ohio and London, England. Forensic journal articles are read in Peoria, Illinois and Pretoria, South Africa. Mass disasters bring forensic experts together from all over the world.

The modern forensic world is expanding. Forensic scientists travel around the world to attend international meetings. Students graduate from forensic science educational programs in record numbers. Forensic literature—articles, books, and reports—grows in size, complexity, and depth.

Forensic science is a unique mix of science, law, and management. It faces challenges like no other discipline. Legal decisions and new laws force forensic science to adapt methods, change protocols, and develop new sciences. The rigors of research and the vagaries of the nature of evidence create vexing problems with complex answers. Greater demand for forensic services pressures managers to do more with resources that are either inadequate or overwhelming. Forensic science is an exciting, multidisciplinary profession with a nearly unlimited set of challenges to be embraced. The profession is also global in scope—whether a forensic scientist works in Chicago or Shanghai, the same challenges are often encountered.

The International Forensic Science Series is intended to embrace those challenges through innovative books that provide reference, learning, and methods. If forensic science is to stand next to biology, chemistry, physics, geology, and the other natural sciences, its practitioners must be able to articulate the fundamental principles and theories of forensic science and not simply follow procedural steps in manuals. Each book broadens forensic knowledge while deepening our understanding of the application of that knowledge. It is an honor to be the editor of the Taylor & Francis International Forensic Science Series of books. I hope you find the series useful and informative.

Preface

The discovery of extensive numbers of polymorphic genetic loci in DNA has revolutionized the way we, as the general public and as scientists, view the role of science in crime scene investigation. We no longer are willing to accept Columbo's clever, last-minute scene revelation of the obscure mental clue that linked the criminal to the crime. Today, we want our DNA. Everyone, from grade school children to judges and juries, expect and demand that when someone is accused of a crime that DNA evidence provide that essential link between the defendant and the victim.

We have made tremendous leaps forward in the last two decades with new DNA technologies that now permit individual identification of source from incredibly small amounts of evidence that may have been deposited many days, months, or even years before recovery by law enforcement. The development of offender DNA databases provides a crime-solving power never before experienced, especially for no-suspect cases. The association of a suspect with the victim or the crime scene through DNA linkage is one of the most powerful statements of complicity in a crime imaginable.

The advent of this unprecedented power to link criminals with crimes has changed the very way we think of guilt or innocence. The sacrosanct standard of criminal culpability of "beyond a reasonable doubt" has mutated into "beyond DNA linkage," for if the evidence includes DNA evidence that associates the defendant to the victim or crime scene, the certainty with which we believe that this defendant committed the crime is almost absolute. The corollary is fortunately, equally true, for if DNA evidence is recovered and provides no link to an accused, the conclusion is one of factual and not simply legal innocence. No category of evidence has ever had the complete capacity to convict or exonerate an accused so absolutely in the minds of the general public. In reality, of course, there are some technical limitations to DNA testing.

Forensic scientists appreciated this awesome power to convict or free individuals early in the genetic revolution. Blood group markers could exclude the accused, but never absolutely and uniquely include a suspect. DNA, with discriminatory powers many orders of magnitude beyond those of blood group markers, provided the necessary evidence of absolute linkage. By 1985, the forensic science community was invested in linking suspects to victims and crime scenes through DNA polymorphisms. The technical difficulty of having large quantity, high quality DNA source material was soon overcome with polymerase chain reaction (PCR) technology. The need for certainty was met by adding more and more genetic loci to the probability statistic until the discrimination power of linking the sample at the scene to one individual (or his evil twin) became probabilistically unique.

Today, armed with genetic evidence from a variety of DNA markers including short tandem repeat (STR) polymorphisms, Y-chromosome STRs, single nucleotide polymorphisms (SNPs), and mitochondrial (mtDNA) markers, forensic science has developed an overwhelming arsenal in the war on crime. But cases still go unsolved. Homicide clearance rates still hover below two thirds, meaning that we have a one in three chance of never solving a murder! The reason for this dismal success rate has many components, but one major component is the lack of DNA evidence to link a suspect to the victim or crime scene. But what if other physical evidence is present, evidence that is not from the suspect, but from the suspect's environment? Here linkage would be strongly circumstantial that the suspect was in contact with the victim or crime scene and transferred evidence from his person to the scene. And if this evidence is biological evidence, can we demonstrate absolute linkage or are we simply limited to mere similarity?

The science of nonhuman DNA typing has made major in-roads for providing this critical link between the suspect and the scene and victim. Polymorphic genetic loci in cats and dogs and other species have been successfully used to provide the association between the human suspect and the human victim. The genetic basis underlying the linkage process is no different from the genetic basis of linking human suspects. Polymorphic genetic loci in nonhuman species can provide the same absolute association, provided that the logical inference of a particular cat or dog can be demonstrated to be in the presence of the suspect or victim. Nonhuman biological material, usually hair, plants, or insects, is transferred from the source to the human to the crime scene following Locard's principle of exchange. Circumstantial evidence of this type is equally probative to the issue at hand, that is, demonstrating the proximate relationship between the suspect and the victim and the crime scene.

Additionally, not all forensic science is directed at solving homicides. Crimes against property, including fraud and contraband, are also rampant. These crimes, while not as publicly visible as homicides, are major drains on society through economic loss, drug addiction, and environmental degradation. Nonhuman DNA once again provides a significant potential to solve these crimes, through linkage of genetic polymorphisms in biological evidence.

The association of large-scale marijuana distribution operations has been well documented. Because marijuana plants are clonally propagated from cuttings from a parent plant, all the siblings share the identical DNA. This fact permits the association of marijuana plants seized in one location to marijuana plants seized in other locations, thereby providing the circumstantial linkage that the groups of illegal contraband are related. Similarly, genetic identification of plant material, including wood chips and leaves, permits the association of crime scene evidence with plants from remote locations, thereby linking suspect and crime scene. The technical accomplishment of demonstrating the genetic association between the plant materials is one of the marvels of modern DNA science. What we have learned over the past two decades is that polymorphic genetic markers abound throughout the genome of all species. With genetic marker determinations through advanced techniques such as AFLP, we can establish genetic identity with similar levels of absoluteness that is seen in the human polymorphic loci identified with STR techniques.

Not all forensic science is devoted to criminal investigation. One of the major impacts of DNA technology is in civil or economic applications. Human paternity testing has become so exact that probabilities are expressed in one in trillions that a purported father is the biological source of half of the child's DNA. The standard human identification STR loci are used to provide the determination of paternity. But what if the question of paternity involves a horse or cow or some other species of great economic value? The answer again lies in the analysis of polymorphic genetic loci that permit the absolute or near absolute determination that a particular animal is the progeny of a particular ancestor. What is different in comparison to human DNA testing is that the genetic markers and the databases necessary to compute the probability of relatedness must be developed anew for each species.

This same technical limitation also applies to applications in which we are interested in simply knowing what species we are examining. Conservation efforts depend on protecting endangered species from eradication by humans interested in the economic value of an exotic rare animal, either alive or dead. The ability to determine the species designation of an item that may no longer morphologically resemble the source, perhaps carved ivory or sewn leather goods, is critical to halting the trade of endangered animal products. The issue is how the genetic markers that uniquely identify one species are established, and how discriminatory are these markers in separating closely related species or subspecies?

Implicit in the nature of forensic science is the fact that the results of all forensic DNA testing are going to be applied in a legal context, that is, testimonial evidence in court. Paralleling the development of new DNA technologies, the development of the law of scientific evidence has undergone radical change in the last several decades. The old standard *Frye* test of general acceptance has been replaced by the *Daubert* standard of reliability. We no longer question whether scientists agree that a particular novel science is sufficiently well accepted that they, as scientists, have faith in the technique. Since 1993, in all federal courts and in most state courts, the question has become not only is the science generally accepted, but also is the science one that follows the principles of the scientific method complete with independent verification of the technique through peer review and determination of the sources of error and error rates. *Daubert* is a threshold question of admissibility for which the human DNA technologies have passed with flying colors. The nonhuman application of the various approaches to polymorphic genetic markers is not so well established. Scientists investigating these techniques should expect rigorous examination of their methodologies, their publications, and their error rates before the admission of nonhuman DNA determination becomes routine.

It should go without saying that the best technology, if improperly applied, is not admissible in a court of law. The responsibility for preparing proper protocols and the exact execution of these protocols is the burden of good science that falls on each of us. Even though we may conduct our testing at the behest of one party or the other, the ethical obligation of being an objective scientist means that we, as scientists, are not advocates. The legal system has evolved over the last thousand years as the most exact truth-finding system ever devised. For proper determination of the truth, it is essential that the scientific facts be presented as scientific facts and facts alone. Let the lawyers advocate, let the scientists be objective.

Albert B. Harper, Ph.D., J.D.

Acknowledgments

Thank you to all the authors named in this book volume for their efforts and contributions to this text and to the field of nonhuman forensic science research. It takes considerable time and effort to edit a text, but even more effort is required to write each chapter and provide high-quality illustrations. This is an important area of study, since the new application of DNA technology to the identification and potential individualization of the innumerable number of plant, animal, and insect species that may be used to associate persons, locations, weapons, or vehicles may be invaluable to solving a case. In addition, a better understanding of genetic diversity in species will aid in breeding programs, wildlife management, and in providing improved random match probability estimates for forensic DNA statistics and population studies.

Thank you to my parents, especially my father, for encouraging me to finish editing this text on an almost weekly basis. A special thanks to my daughters and husband for understanding the time away from them to complete this book. Thanks to the forensic science faculty at the University of New Haven, in particular Dr. Henry C. Lee, Prof. Timothy M. Palmbach, Dr. Albert Harper, and Dr. Howard Harris who I hope will consider using this as a text for their curriculum and evidence training workshops. Several University of New Haven students are named in this text, assisted with this book and their efforts are greatly appreciated. Finally, last but certainly not least, thanks to Becky McEldowney and Kari Budyk at CRC Press for their great patience and assistance with publishing this text. I hope the readers will find it interesting and useful when considering the added value of nonhuman forensic evidence at the crime scene, in the laboratory, and in the courts to solve crimes.

Editor

Dr. Heather Miller Coyle is an assistant professor of forensic science at the Henry C. Lee College of Criminal Justice and Forensic Sciences at the University of New Haven. Dr. Miller Coyle teaches introductory forensic science in both the undergraduate and graduate forensic science programs at the university as well as being an instructor for a variety of training workshops for the Henry C. Lee Institute. She specializes in DNA testing for the individualization of human and nonhuman evidence; her research interests include forensic botany and the development and application of new DNA techniques for highly challenging samples such as minute plant materials, grass stains, seeds, DNA from fingerprints, and bone samples from exhumed and mummified remains.

Dr. Coyle received her Ph.D. in plant biology from the University of New Hampshire in 1994 and performed postdoctoral DNA research at both Yale University and Boehringer Ingelheim Pharmaceuticals, Inc. She has lectured on forensics for numerous professional organizations and has taught forensic DNA courses at Wesleyan University and the University of Connecticut. Dr. Coyle also edited and authored a companion text entitled *Forensic Botany: Principles and Applications to Criminal Casework*, CRC Press, in 2005. She has published many scientific articles in professional journals and contributed book chapters on basic DNA methods and plant DNA research. At the undergraduate and graduate levels, Dr. Coyle serves as a scientific advisor for students beginning their careers in the forensic sciences to assist them with novel research and scientific publication. She also has testified as an expert witness for criminal cases involving DNA identification and reviews both criminal and civil casework for scientific integrity.

Contributors

Khudooma Said Al Na'imi, B.Sc
Forensic Science Laboratory
Abu Dhabi Police G.H.Q.
Al Ain City, Abu Dhabi, United Arab Emirates

Burkhard Berger, Ph.D.
Institute of Legal Medicine
Innsbruck Medical University
Innsbruck, Austria

Peter Bilous, Ph.D.
Department of Chemistry/Biochemistry
Eastern Washington University
Cheney, Washington

Victor A. David, Ph.D.
Laboratory of Genomic Diversity
National Cancer Institute at Frederick
Frederick, Maryland

Pero Dimsoski, Ph.D.
Genoma
Menlo Park, California

Cordula Eichmann, Ph.D.
Institute of Legal Medicine
Innsbruck Medical University
Innsbruck, Austria

Dr. Paolo Garofano, M.D., Ph.D.
Forensic Genetics Division
UNIRELAB s.r.l
Milan, Italy

L.H.P. van de Goor
Dr. van Haeringen Laboratorium BV
Wageningen, Netherlands

Joerg Graf
Department of Molecular and Cell Biology
The University of Connecticut
Storrs, Connecticut

Dr. W.A. van Haeringen, Ph.D.
Dr. van Haeringen Laboratorium BV
Wageningen, Netherlands

Albert B. Harper, Ph.D., J.D.
Henry C. Lee Institute of Forensic Science
West Haven, Connecticut

Jonathan Hill
Department of Molecular and Cell Biology
The University of Connecticut
Storrs, Connecticut

En-Uei Jiang
Forensic DNA Laboratory
Ministry of Justice Investigation Bureau (MJIB)
Taiwan, R.O.C.

Richard M. Jobin, Ph.D.
Alberta Sustainable Resource Development
Fish and Wildlife Forensic Laboratory
Edmonton Alberta

Henry C. Lee, Ph.D.
Forensic Science Program
Henry C. Lee College of Criminal Justice and Forensic Sciences
University of New Haven
West Haven, Connecticut

Holly Long, B.S.
Honors Thesis Program
Forensic Science Program
Henry C. Lee College of Criminal Justice and Forensic Sciences
University of New Haven
West Haven, Connecticut

Marilyn A. Menotti-Raymond, Ph.D.
Laboratory of Genomic Diversity
National Cancer Institute at Frederick
Frederick, Maryland

Stephen J. O'Brien, PhD.
Laboratory of Genomic Diversity
National Cancer Institute at Frederick
Frederick, Maryland

Walther Parson, Ph.D.
Institute of Legal Medicine
Innsbruck Medical University
Innsbruck, Austria

John Schienman, Ph.D.
Center for Applied Genetics and Technology
University of Connecticut
Storrs, Connecticut

Linda Strausbaugh
Center for Applied Genetics and Technology
Laboratory for Non-Traditional DNA Typing
The University of Connecticut
Storrs, Connecticut

An Introduction to Forensic Science and DNA

1

HEATHER MILLER COYLE, PH.D.

Contents

1.1 Forensics Defined

Forensic science is a compilation of scientific and analytical methods "borrowed" from multiple disciplines and applied to matters of law. At the most simplistic level, forensic science is about performing reliable tests on evidence properly collected from crime scenes to aid in case resolution.[1] By collecting physical, chemical, or biological evidence, extensively documenting the evidence at both the scene and at the laboratory, and performing analyses using various laboratory tests, crime scene professionals are often able to assemble an amazingly detailed account of the crime (reconstruction) (Figure 1).[2] Those individuals that perform crime scene reconstruction must know most aspects of crime scene and laboratory analyses to give a probable explanation of events. Although most people think of famous homicide cases when they consider forensic science, forensic science actually encompasses both criminal and civil casework.

The ability to associate items of evidence to a scene or person is based on a principle called "Locard's Exchange Principle," which states that when two objects come in contact, they exchange or transfer small amounts of information to each other.[1–4] Once this transfer is detected and the substance classified and/or individualized, the forensic scientist will have a clue as to what may have occurred at the scene. Although there are many different

Figure 1 The basic principles of evidence recognition, collection, and preservation are the critical first steps in ultimately being able to utilize nonhuman DNA typing methods in forensics. At the examination step, the criminalist will review the documentation, collection, and preservation procedures before beginning testing of the evidence. Photograph: Mr. John Doyle, University of New Haven.

subspecialties in forensic science (e.g., document examination, shooting reconstruction, fingerprints, firearms, serology, and more), this book will focus on methods for the identification of biological substances for forensic applications. This first chapter is designed to present a foundation for the further discussion of DNA typing not just from humans, but also from other biological organisms. DNA typing of other organisms such as plants,[5] animals, insects, bacteria, and viruses will play an important role in forensic science,[6,7] especially as we enter a time of increased bioterrorism and mass attacks on human civilization. In order to better prepare for such events, nonhuman DNA tests need to be identified, further developed for forensic science, and presented in courts of law.

Human identification tests are useful in ascertaining the identity of a body or the depositor of biological evidence. These tests include identification and individualization of a person by fingerprint ridge patterns or by analysis of DNA fragment patterns. The detailed analyses of fingerprint ridges and the patterns left by secretions from the sweat glands that line them provide valuable clues as to the identity of a victim, witness, or the perpetrator of the crime. Everyone, including identical twins, can be uniquely identified by their fingerprints. While identification by fingerprint analysis is infinitely useful, DNA-based identification techniques have also come to the forefront of forensic science (in the past 10 years), especially for determining the depositor of biological stains such as blood, saliva, semen, urine, and sweat from a wide variety of objects.[2,8,9] Due to the increased use of DNA for human identification, numerous crimes from both new and "cold" cases have been solved.

The technology for using DNA to identify individuals (everyone except identical twins have unique DNA profiles) has evolved at a rapid pace and, to date, most forensic laboratories are performing DNA typing with the same core set of DNA markers. These common markers are provided in standardized commercial test kits and include 13 short tandem repeat (STR) tests combined into a single tube for high throughput processing. The use of a core set of markers facilitates searching a national forensic DNA database to identify potential contributors of the biological evidence to the crime scene.

1.2 What is DNA?

DNA, deoxyribonucleic acid, is a macromolecule found in the nucleus of all living cells and contains the genetic information that is required to "type" or identify an individual. Nuclear or "genomic" DNA is a double-stranded molecule organized into condensed packages called chromosomes. In humans, there are 23 pairs of chromosomes that house billions of base-pair units of DNA. A large percentage of that DNA is conserved, or similar, in humans as it is required to code for the proteins that make us who we are. However, there are segments of DNA that are known to be different from person to person within any given population and those are called "hypervariable" regions. These segments of DNA that differ between individuals have been selected and optimized as markers for human identification.

Returning to the compressed units of DNA called chromosomes, one half of each chromosome pair is inherited maternally and the other is inherited paternally. This pattern of inheritance is used to establish parental and sibling relationships when identifying missing persons or accident victims. When a chromosome is untangled from the associated proteins (e.g., histones) that help hold its shape, it can be visualized as a twisted ladder in which the most basic unit is called a nucleotide. Nucleotides are linked together to form both the sides and rungs of the molecular ladder and each nucleotide can be broken down into three parts: a nucleotide base, a sugar, and a phosphate. The nucleotide bases create the rungs of the ladder by forming chemical bonds. These bases (adenine [A], thymine [T], cytosine [C], and guanine [G]) always bind in a predictable fashion (i.e., complementary base pairing) such that A binds with T and G binds with C. The sequence of these nucleotide bases provides the critical information called the *genetic code* and individualizes each living organism.[3,4]

Currently, most forensic laboratories are performing nuclear-DNA-based tests that analyze segments of variable DNA called *short tandem repeats*, abbreviated as STRs. These segments of DNA vary in length and are made up of a repeating series of four nucleotide base units. An example of this would be a repeating sequence of AGGT-AGGT-AGGT, which would be designated as a 3′ for that segment of DNA for that individual's profile. Thirteen segments or loci are tested and the band patterns that are generated are converted to numeric values for ease of comparing the evidentiary profile to the known reference profile.

Another form of DNA, mitochondrial DNA (mtDNA), can be found in subcellular organelles called *mitochondria*. The human mitochondrial genome is a circular molecule that consists of 16,569 nucleotides and is maternally inherited. Since the mtDNA molecule and all of its DNA information is passed from mother to daughters and sons, it can be used to associate family members that share a common maternal lineage. This is particularly useful for associating family members from missing person cases or mass disasters. Mitochondrial DNA analysis is used when sample quantities are extremely limited, if a nuclear DNA test results fail, if cell sources are ancient, or if the source does not contain nucleated cells (e.g., a hair shaft). Since all maternal relatives share the same mitochondrial DNA type, mtDNA analysis is inherently not as discriminating as genomic DNA analysis but has the significant advantage of requiring less DNA than many other types of tests. For mitochondrial DNA analysis, two regions that vary between individuals within populations are tested for differences in the sequence of the nucleotide bases. As with other types of DNA testing, a known reference sample is required to compare the DNA test result against the evidence.

Sex chromosomes (designated X, Y) are useful in forensics for establishing whether the depositor of the biological sample is male or female. Incorporated into most commercial kits are sex-specific DNA markers for this purpose; commercial kits are also available for DNA markers located exclusively on the male chromosome. These markers are called Y-STRs (Y chromosome short tandem repeats) and are used to generate a male-specific profile. Y-STR typing is valuable for separating out the male component in male-female mixtures that commonly occurs in sexual assault casework (female epithelial cells and male sperm or epithelial cells). Due to the acidic vaginal environment, cells begin to degrade even before they can be collected for analysis. Although forensic DNA extraction procedures attempt to separate out the different cell types (epithelial and spermatozoan), it is not always possible to separate the DNA if the cells have broken open and their DNA contents have become mixed. Y-STR typing is a good option for separating the DNA profiles based on sex of the contributor and can establish if more than one male contributor is present in the DNA mixture. Y chromosome testing is used in sexual assault cases, paternity cases, mixed biological sample cases, and as a screening tool for sorting out the possible number of contributors to stains on a garment prior to additional nuclear DNA tests. While there are many benefits to Y-STR tests, the genetic variation that exists in the male population with current DNA markers is not as great as with nuclear STR tests; therefore, some unrelated males will have the same Y-STR type.

Virtually all of the forensic DNA tests available today use a molecular copying process called polymerase chain reaction (PCR), which can take a limited quantity of DNA sample (1–10 nanograms) and expand the amount to a detectable level to generate a DNA profile (Figure 2). This process involves the synthetic replication of DNA under controlled conditions using enzymatic reactions in a tube to mimic the replication of DNA when cells divide in the human body. During the copying process, fluorescent dyes are incorporated into each copy of DNA so that the DNA can be later visualized as a separate DNA fragment on a DNA sequencer, analogous to a colored bar code. The DNA fragments are separated by fragment size and molecular charge in a gel polymer matrix using electrophoresis on the DNA sequencer instrument and visualized as they pass by a laser. The laser excites the fluorescent dye on each fragment and a CCD camera records the time and location of the band to later reconstruct a computerized image of the DNA profile. Computer software then enables the DNA analyst to assign number values to the DNA fragments to generate the DNA profile that can be compared to the known reference sample(Figure 3). Regardless of whether the test is for DNA fragment length analysis (STR markers) or for DNA sequencing (mtDNA tests), the PCR process is used to first expand the sample so that sufficient quantity is available for testing.

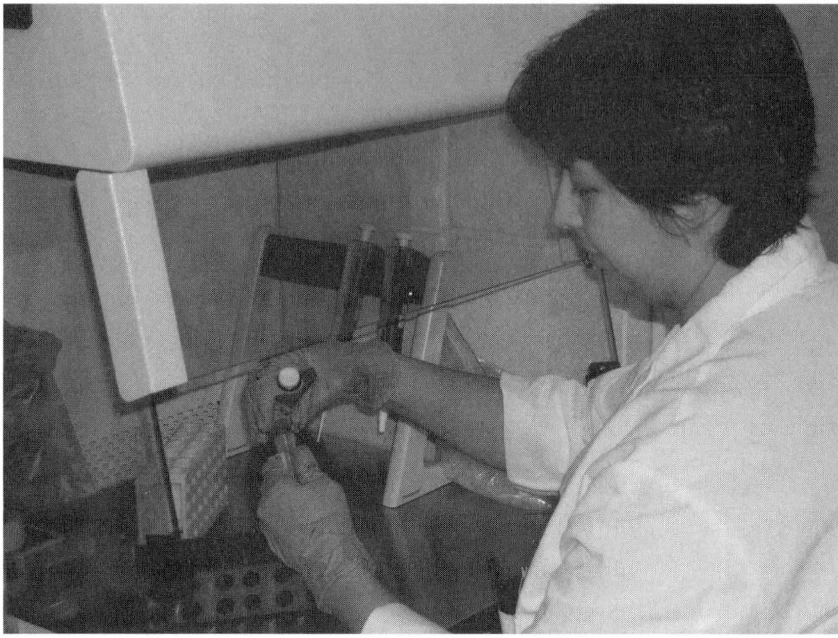

Figure 2 PCR, or polymerase chain reaction, is an enzymatic reaction in which DNA is "copied" from one molecule into billions of exact copies, thereby allowing for a single cell to be detected from a crime scene. Since the PCR reaction is sensitive enough to allow for copying DNA from just a single cell, the work is performed in a laminar flow hood to prevent possible DNA contamination of the sample from the DNA analyst. Photograph: Ms. Jennifer Nollkamper, University of New Haven.

1.3 Benefits and Limitations of DNA

Validation experiments have shown that nuclear DNA is the same in every somatic (nonreproductive) cell within an organism, which is useful in the sense that DNA test results can be appropriately compared from blood left at a crime scene to epithelial cells collected as the known reference sample. DNA profiles also cannot be altered by environmental factors so that one person's DNA profile could be changed to look like another person's. With the exception of identical twins who share identical DNA profiles, each person's DNA is unique and can be used as an identifier. On occasion, blood transfusions can result in mixtures of two DNA profiles because the DNA profile of both the donor and recipient is detected. While oftentimes a DNA profile can be generated from forensic evidence, several factors affect the ability to obtain a DNA profile in entirety and sometimes a DNA profile is not possible to obtain. The first factor is sample quantity. On average, 1 nanogram

Figure 3 Most forensic science laboratories have sophisticated DNA sequencers that will detect DNA fragments that have been dye-labeled and separated on a gel matrix during electrophoresis. Computer software will then convert the complex band patterns into numeric code so that DNA profiles may be easily stored and compared. Photograph: Ms. Melissa Benevides, University of New Haven.

of DNA is necessary to generate a good DNA profile but sometimes much less (picograms) is sufficient for DNA typing. Secondly, a sufficient sample quality is required, and high molecular weight DNA is good for generating full DNA profiles. The third consideration is sample purity. STR DNA typing methods are typically not as affected by dirt, grease, fabric dyes, and leather tannins as were some of the earlier DNA typing methods.[3,4]

1.4 DNA Databases

There are three types of DNA databases (sometimes called databanks): reference population databases, convicted offender databases, and unsolved crime-scene DNA databases (often referred to as *no-suspect cases*). Reference population databases are random samplings of various populations to determine if there are any differences in allele or DNA marker frequencies that would affect the prediction of how often one might expect to see a DNA profile (i.e., genotype) on a particular piece of evidence. A statistical evaluation of these databases can establish if there are common types or rare types in a population and can be used to give weight or meaning to a match between DNA from evidence and a known reference sample. Convicted offender databases are collections of DNA profiles that have been obtained by court order from felons that have been previously convicted of crimes such as violent homicides or sexual offenses. Each state has its own legislation that dictates from whom DNA samples may be collected. Access to the DNA databases is limited to law enforcement and participating forensic laboratories, and the information contained within them is not for dissemination to the general public. Each sample has a coded letter designation as a reference to the sample and can be cross-referenced to DNA profiles obtained from probative evidence from unsolved crimes. The third type of DNA database is a collection of these no-suspect DNA profiles from crime scenes that can be searched to determine if any of the DNA profiles can be used to link different cases together. They also can be searched against the state and federal convicted offender databases to establish if they match a previously convicted offender profile and, thus, are useful in providing investigative leads and potentially solving cases.[3,4]

As the field of forensic human DNA test methods expands into that for nonhuman DNA testing, the development of appropriate reference population databases will be essential for each new test method and the organism to be tested. To generate a relevant DNA database, some knowledge of the genetic history and reproductive strategies of the organism will be essential so that sampling issues can be addressed during the construction of the database and during later court testimony.[5–7]

1.5 The Future of Forensic DNA

As the human DNA typing methods for forensic individualization have become more uniform from laboratory to laboratory, the technology has finally stabilized. Initially, as methods were being developed, new DNA test methods were being introduced and implemented every few years. Now, almost all forensic testing laboratories in the United States use the Combined DNA Indexing System (CODIS) core loci, which has allowed for the construction of a federal DNA database of convicted offenders. Where does the field of forensic DNA typing go from here? As diverse as human beings are in regards to physical traits and cultural heritage, they still represent only one species. There are literally hundreds of thousands of other biological species that can be useful as potentially probative forensic evidence. The major classifications include plants, animals, bacteria, viruses, and insects, and the subsequent chapters in this textbook will describe in greater detail how to collect and preserve these forms of evidence. In addition, each chapter contributor is a leader in new applications of forensic identification and individualization and they have provided excellent descriptions of the state-of-the-art testing technology in their area of expertise. Along with the test methods for nonhuman DNA, also will come the need for the construction of representative species databases for legal acceptance as the new technology becomes implemented and accepted into the courts.

References

1. Stuart, J.H. and Jon, N.J., *Forensic Science: An Introduction to Scientific and Investigative Techniques,* CRC Press, Boca Raton, FL, 2003.
2. Lee, H.C., *Physical Evidence,* Magnani and McCormic, Enfield, CT, 1995.
3. Butler, J., *Forensic DNA Typing,* Academic Press, San Diego, CA, 2001.
4. Inman, K. and Rudin, N., *An Introduction to Forensic DNA Analysis,* CRC Press, Boca Raton, FL, 1997.
5. Miller Coyle, H., *Forensic Botany: Principles and Applications to Criminal Casework,* CRC Press, Boca Raton, FL, 2005.
6. Day, A., Nonhuman DNA testing increases DNA's power to identify and convict criminals, *Silent Wit.*, 6, 2001.
7. Jobin, R., Patterson, D.K., and Stang, C., Forensic DNA typing in several big game animals in the province of Alberta, *Can. Soc. Forens. Sci. J.*, 36, 56, 2003.
8. DeForest, P.R., Gaensslen, R.E., and Lee, H.C., *Forensic Science: An Introduction to Criminalistics,* McGraw-Hill, New York, 1983.
9. Gaensslen, R.E., *Sourcebook in Forensic Serology, Immunology and Biochemistry,* U.S. Government Printing Office, Washington, D.C., 1983.

Collection and Preservation of Biological Evidence from the Crime Scene

2

HENRY C. LEE, PH.D.

Contents

2.1 Recognition of the Evidence

One of the most difficult aspects of forensic investigation comes first, the crime scene. If evidence is not recognized as being probative or important to the case, if it is not documented and collected, then the evidence will not be available for subsequent laboratory testing.[1] A good rule of thumb is "when in doubt; collect it," as it may end up being a crucial piece of information. Recognition of evidence comes with training and experience in forensic investigation. Physical evidence is required to help prove or disprove certain facts in a criminal case; without it, there may be little to support or refute a victim or suspect's statements of the crime. Fortunately, there are many new developments for searching a crime scene and examining evidence, including tactile and visual searches frequently aided by use of alternate light sources or chemical tests to detect stains, fibers, or chemical compounds. Although there is a tremendous burden on laboratories to process evidence as speedily as possible, great care should be taken to examine and document each piece of physical evidence found at the crime scene and in the laboratory to avoid later repercussions in court.

Standard evidence collection guidelines as applied to human biological sources and trace evidence can be extended to samples from nonhuman sources as well. Nonhuman sources of biological evidence could include animal blood, tissue, and hair; insects;[2] plants and other vegetative matter;[3–5] viruses; and bacteria.[6] As we anticipate and plan for future attacks by bioterrorists and criminals, an increase in the utilization of nonhuman biological evidence is a natural extension of current forensic capabilities. Even now, when the evidence is limited, any and all forensic tools are useful for achieving case resolution. Since its inception, forensic science has been a compilation of scientific disciplines in which experimental techniques have been applied and scientifically tested (validated) to define the benefits and limitations of each method. Techniques that are widely applicable to forensic evidence and meet court standards become used routinely in forensic casework. The highly variable nature of crime scenes presents new challenges and new techniques, which further enhances the ability to solve crimes.

2.2 Fundamentals of a Crime Scene Search

The foundation of any investigation rests on the crime scene. Therefore, it is extremely important to follow standard crime scene procedures. Established priorities at the crime scene include the following:

1. Save and preserve life and render medical assistance if necessary.
2. Detain any witnesses or apprehend any suspects that may still be in the vicinity.
3. Secure the crime scene and limit any unnecessary personnel or traffic so as to preserve the condition of the scene and any relevant evidence.
4. Survey the crime scene for any transient, conditional pattern or associated evidence.
5. Document, collect, and preserve physical evidence from the crime scene.

Since crime scenes vary in condition and are dependent on the type of crime, searching for physical evidence may require multiple search patterns and different techniques. The type and condition of the crime scene may dictate the order in which the search occurs. The order may also depend on any transient conditions at the crime scene. For example, outdoor areas may be processed first due to environmental or weather factors that could quickly alter the evidence. A body may be processed next, since samples, especially DNA samples, may deteriorate over time if contained within body cavities or on the surface of the body. Indoor crime scenes may be processed last; the general rule is to protect, preserve, and record any physical evidence *in situ* according to standard practices.

2.3 Biological Evidence Collection and Preservation Procedures

2.3.1 Blood

Liquid blood from any source, human or animal, should be collected with sterile swabs or gauze and allowed to air dry at room temperature. Swab collection boxes or clean, dry paper envelopes may be used as containers for the dried swabs. It is important to keep the sample out of direct sunlight while drying and let it dry completely, and especially to be aware of humid climates that may delay the drying process. Once dry, swabs can be refrigerated or frozen prior to processing, but freeze-thaw cycles should be avoided. Crime scene personnel should protect themselves with coveralls, boots, and disposable gloves, since liquid blood may contain infectious diseases such as hepatitis and human immunodeficiency virus. When evidence is repackaged, it should be noted on the package and documentation worksheets so that there is a clear record of chain of custody.

A dried bloodstain that is already adherent to clothing or another object should be allowed to remain in the dry state. Clothing should be wrapped in clean, dry paper and placed in a brown paper bag and sealed. No attempt to cut out stains from the clothing should be made until the item has been fully photographed and documented, since the location, the condition, and the quantity of the stains may be useful to the case for later crime scene reconstruction. At the forensic science laboratory, cuttings from clothing where stains have been identified may be taken and analyzed further.

Small objects can be wrapped, secured, and sent directly to the laboratory. When it is impractical to send large items to a laboratory, the bloodstain may be scraped into a paper fold with a disposable scalpel and sealed in an envelope. Also, the area could be swabbed or the blood transferred onto a sterile swab prior to being placed in an envelope. Another option is to lift the dried bloodstains or crusts off the immovable object using fingerprint tape or gel lifters. Some surfaces may not release bloodstains easily, so it is a good idea to perform a test lift before trying to remove the sample.

2.3.2 Other Body Fluids

Other body fluids may fluoresce under different wavelengths of alternate forensic light sources, which can aid the investigator or scientist in identifying possible stains for further testing. These body fluids may include both human and animal sources of semen, saliva, urine, sweat, tears, and vaginal secretions. Many of these body fluids have both presumptive (preliminary) and confirmatory (additional) tests; several only have presumptive tests that have been developed and are in standard use at forensic laboratories. Collection methods for these body fluids parallel the methods described above for liquid and dried blood.

The investigator or scientist should keep in mind that evidence recognition and collection are two of the most difficult and critical steps in a forensic investigation. It is possible that a stain on a sheet that fluoresces at the same wavelength of light as semen will not actually be semen (e.g., a false positive such as bleach). False positives are common with many preliminary chemical test methods conducted in the field and, therefore, additional testing must be performed in the forensic laboratory in order to reach a definitive identification of a particular body fluid. Some soil components and plant enzymes can give false positives in presumptive blood identification tests, so outdoor crime scenes, clothing, and shoes may be affected. False negatives are also possible in forensic field testing. A false negative might be expected if a substance associated with the body fluid inhibits the chemical reaction required to give a color change in the body fluid presumptive identification test. Therefore, any suspected biological stain should be collected and submitted to the laboratory for further analysis.

2.3.3 Hair and Other Biological Trace Materials

Hair and biological trace materials, whether of human or animal origin, are typically collected with sterile or disposable tweezers, placed in a paper fold, and sealed in an envelope for later analysis. This is important if the hair is visible (e.g., in the hands of a victim) and likely to be lost in transport. These types of hair samples should be photographed and collected as soon as possible. Frequently, clothing, small objects, furniture, weapons, and cuttings from carpets will be packaged and documented at the crime scene. After these items reach the forensic science laboratory, they will be examined and scraped for collection of trace materials. Scraping of an item involves shaking of the item over a large, clean sheet of paper to collect nonadherent debris such as soil particles, hairs, fibers, plants, seeds, pollen, and insects. Subsequently, the analyst will use a straight edge tool to physically scrape off any adherent items for collection and analysis. If the item is not excessively large, a thorough examination with an alternate forensic light source and collection of hairs, fibers, plants, and seeds, plus animal and insect parts with tweezers are also options.

For human and animal hair samples, a microscopic examination of the hair(s) and comparison to known reference samples will determine if the known reference person or animal could have been the source of the evidentiary hair. If the conclusion is that the known reference is consistent with being a possible source, then further DNA analysis may be conducted. Nuclear DNA analysis is used to test tissue from the root area of the hair, but broken hair shafts or telogen hairs require mitochondrial DNA testing. Microscopic identification and subsequent DNA analysis of other animal and plant material may also reveal possible leads for further investigation.

2.3.4 Tissues, Bones, Teeth

When bones or teeth are discovered, a forensic anthropologist or odontologist may be called in to determine the origin of the sample. Many animal bones resemble human bones, and it is difficult for a nonexpert to discern the differences between the two sources. After photographic documentation, old or dried bones and teeth should be placed in clean, dry paper and stored in breathable boxes or paper envelopes for later analysis and DNA testing. The source of cells in old bones and teeth are the center marrow and dental pulp, respectively. Therefore, long bones with greater quantities of marrow are preferred. Also, intact bones and teeth are better sources of DNA, as they have essentially served as protected containers for the marrow and pulp and are less likely to show signs of DNA contamination. The outer surfaces of the bones or teeth should not be cleaned prior to their reaching the forensic science laboratory, as this may prevent good DNA test results, particularly if the samples are fragmented.

If tissue is still attached, then the collection and storage methods may be different. When in doubt, contact a forensic anthropologist or the forensic science laboratory for specific instruction. Soft tissue samples are typically frozen and maintained in a frozen condition for later DNA or toxicological testing. Use of chemical fixatives such as formaldehyde is discouraged, as this may inhibit the ability to generate a DNA profile from the tissue.

2.3.5 A Multidisciplinary Approach in Identification of Remains

Cause and manner of death of a human or other animal can be determined by a medical examiner, forensic pathologist, or veterinarian. An examination of injuries or wound patterns on a body can demonstrate the means and facts of the death; defensive wounds, for example, indicate signs of a struggle. Forensic anthropology, forensic odontology, and criminalistics are now used widely in conjunction with DNA analysis to solve a variety of crimes, including cases resulting from war or mass disasters. The identification of a human body involves many steps, including looking at the following identifying features:

1. Dental evidence such as fillings, missing teeth, and tooth repair
2. Physical description of the body
3. Clothing, shoes, jewelry, or other personal effects
4. Documents such as credit cards, driver's license, passport, and ID cards
5. Fingerprints and palm prints
6. Visual identification by a family member or friends
7. Unusual features from medical records (e.g., artificial body parts, implants, pacemakers, prescription lenses, scars, birthmarks)
8. Anthropological assessment of sex, race, and age from skeletal remains
9. Distinctive tattoos, birth marks, and surgical marks
10. DNA identification
11. X-rays, standard medical records

Identification by DNA testing is an expensive and time-consuming venture. Therefore, this test is typically reserved for cases in which other forms of identification have failed or are in dispute. It may not be possible to achieve a DNA test result on a sample that is small or in poor condition. The ability to obtain a DNA profile is dependent on many factors, including the quantity and quality of the biological material available.

2.4 Importance of Nonhuman Biological Evidence in Investigations

Other associative biological evidence may provide a link to the original crime scene, victims, or to a suspect (e.g., pollen, plant material, insects, natural fibers). The value of these forms of nonhuman biological evidence increases in cases where little other traditional physical evidence is available. Numerous cases have been documented in which botanical (plant) and entomological (insect) evidence has proved essential to solving the case.

Case Example:

Lisa, a seven-year-old girl, was reported missing by her mother. An intensive search was conducted by police and other volunteers. The victim's body was found by a firefighter in a grassy field not far from her house. The girl's body was found face down and her clothing located approximately 20 feet from her body. A detective from Milford Police Department contacted

Dr. Henry Lee from the Connecticut State Police Forensic Laboratory to assist in the investigation. The victim was found to have been sexually assaulted and died of manual strangulation. A close examination of her body with an alternate forensic light source revealed several foreign hairs on her body. The microscopic examination identified the hairs as dual-tone, white and black canine hairs. Plant seeds and soil samples were collected as evidence in addition to several control standards for comparison.

During a laboratory examination of the victim's clothing, additional dog hairs were recovered. Microscopical examination revealed the dog hairs could have originated from four different sources. The vaginal swabs collected from the medical examiner's office revealed the presence of spermatozoa. Serological testing showed the donor of the semen was ABO type B, a secretor, and isoenzyme type PGM 1-1. A geographical analysis and crime scene reconstruction was conducted. The preliminary conclusion indicated that the likely suspect was a male with blood type B, living in close proximity of the neighborhood, owner of at least four dogs, and possibly knew or was related to the victim. With subsequent investigation, detectives developed a potential suspect. The suspect was the uncle of the victim and lived two doors from the victim's home. Detectives executed a search warrant and found the suspect did own four dogs named Pepper Snow, Shooter, and Bingo. Known reference hair samples from the four dogs were collected. Microscopical comparison of the hair samples from the victim's body and clothing showed similar characteristics for the medulla and cuticle patterns, color, pigment granules, texture, diameter, and root appearance. These tests showed that the dog hair samples recovered from the victim could have come from the dogs owned by the suspect.

The suspect's clothing and shoes were seized and examined. Grass seed and grass fragments were found on his boots (Figure 1). The plant material was similar to that collected from the scene (Figure 2). After further examination, a botanical expert was consulted and he concluded that both grass samples had the same origin. A comparison of test results from the biological samples from the scene matched to that of the known reference samples for the suspect (Figure 3). The suspect was ultimately tried and convicted of the crime and received a life sentence. This case demonstrates the importance of recognition, collection, and analysis of nonhuman biological evidence. The subsequent chapters in this book will address the specific laboratory testing of many of these forms of evidence and give additional detailed case examples. Several of these chapters will describe animal hair testing and its applications to criminal cases and enforcement of wildlife regulations.

Collection of nonhuman evidence is critical in that many valuable forms such as pollen and bacteria may be difficult to see. For microscopic evidence, vacuuming or tape-lifting may be the only alternatives to ensure that the evidence has not been overlooked. In addition, specialists may need to be consulted when a human or animal bone fragment is found. For cases in

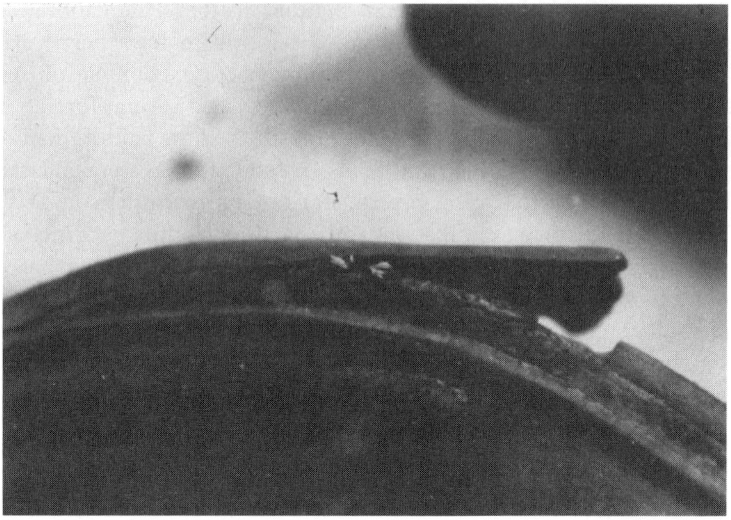

Figure 1 Some grass seeds (located in the center of the photo, bottom of boot) that were recovered from the suspect's shoes ultimately matched back to those collected from the crime scene and proved to be valuable associative evidence. (**See color insert after p. 108.**)

Figure 2 Plant material that was collected from the crime scene is shown in this photo. (**See color insert after p. 108.**) Seed pods are visible and the seeds from the suspect's boots were consistent with coming from the same source plant.

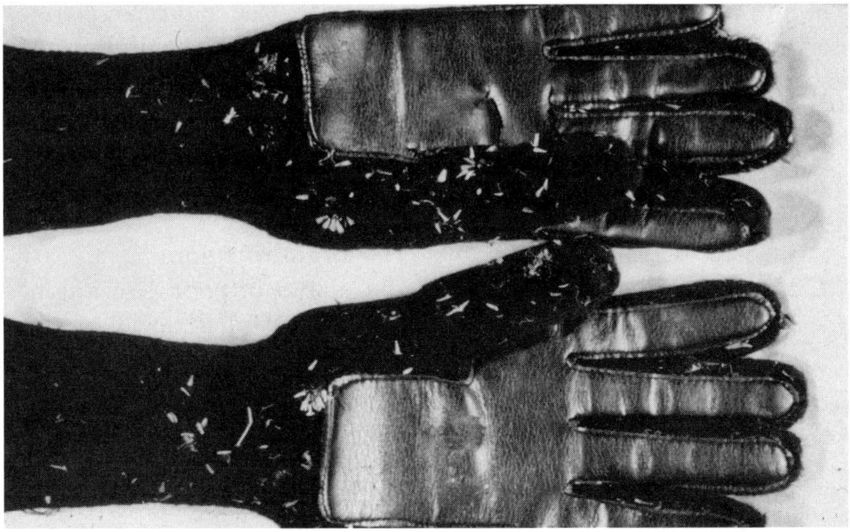

Figure 3 The suspect's gloves collected from the crime scene and sent to the forensic science laboratory were covered in numerous grass seeds. (**See color insert after p. 108.**) A simple microscopic comparison can add extra information to show a linkage between individuals and the crime scene.

which bioterrorism is a possibility, the rapid and precise evaluation of a substance is essential, as specialized teams may be required for management of the scene to preserve the health and safety of personnel. A teamwork approach that relies on expert evaluation when necessary can help preserve the integrity of the crime scene and determine which evidence is best to collect.

2.5 Chain of Custody

The chain of custody refers to the records and documentation of the physical evidence from the time it is recognized and collected at the crime scene to the time it appears in the courtroom. The log of evidence begins at the crime scene and includes:

1. Where it was taken from (location)
2. Who collected it
3. How it was packaged
4. When it was collected
5. The description of the item

From that point on, the evidence must be tracked and accounted for to preserve the chain of custody. If there is a break in the chain of custody, it may be unclear as to where the evidence was and who was in possession of it. This naturally leads to questions of possible evidence tampering when the evidence is introduced at trial. Examples of errors in chain of custody may include but are not be limited to the following:

1. Incorrect labeling or a complete lack of information
2. Insufficient documentation (e.g., no photos or poor description)
3. Failure to note disposition or consumption of the evidence during laboratory testing or transport to and from the laboratory or any other facility
4. Failure to document prior and after removing a sample for testing
5. Failure to note storage conditions at the courthouse
6. Failure to note that an object has been repackaged over time
7. Failure to note the type of test performed
8. Failure to document the results of the testing

Documentation is absolutely critical for later case reconstruction. It is difficult to determine ahead of time which cases may require reconstruction; therefore, a sufficient number of high-quality photos, graphs, and diagrams of measurements that adequately represent the crime scene and crime scene elements are invaluable for later crime scene reconstruction. Even if a biological sample was not collected, the pattern of this stain can be determined from good quality photographs. A scale should be included in the photographs as a size reference. If a ruler is not available, then a coin can be used for relative scale of an object. In particular, blood patterns should be noted and well documented, as they may be useful to later interpret the type, source, and condition of the bloodstains.

2.6 Reexamination of Physical Evidence in Cold Cases

Cold or unsolved cases require the reexamination and reanalysis of physical evidence. Whenever new technology or new information suggests that an older, unsolved case may be resolved, it is useful to reconsider the facts and physical evidence in the case. When performing the reevaluation, however, it is important to consider the forensic context. Thirty years ago, forensic DNA analysis was not available and forensic investigators and examiners may have not used the same precautions that we use in handling evidence today. Therefore, it is possible to obtain a DNA profile on a substance that may not be associated with the crime but be due to subsequent handling practices or cross-contamination from poor storage conditions or examination procedures. A careful and considerate evaluation of these factors should be made prior to performing a DNA test or interpreting the DNA results on cold-case evidence.

2.7 Summary

With the advent of newer, sensitive testing techniques such as DNA testing, however, comes great power and responsibility. New technological advances can assist in solving old crimes, exonerating the wrongfully convicted as well as leading to the true perpetrator of the criminal act. The correct interpretation of the laboratory test results in light of the crime scene and all subsequent actions associated with the evidence require adequate training and experience by all personnel involved with the case. This includes not only the police, detectives, crime scene investigators, and forensic scientists but also the judges, attorneys, courtroom clerks, and evidence technicians who may handle and assess the evidence. Finally, scientific experts in their field have an ethical responsibility to convey accurately to the triers of fact (the judge and jury) the exact procedures and results, so the appropriate conclusion regarding the case can be reached.

References

1. Lee, H.C., Palmbach, T.M., and Miller, M.T., *Henry Lee's Crime Scene Handbook*, Academic Press, Boston, 2001.
2. Halverson, J.L. and Basten, C., Forensic DNA identification of animal-derived trace evidence: tools for linking victims and suspects, *Croat. Med. J.*, 46, 598–605, 2005.
3. Mildenhall, D.C., A grain of evidence: invisible pollen grains can be enough to sort out the guilty and the innocent, *N.Z. Sci. Mon.*, June 1998.
4. Miller Coyle, H., Lee, C.L., Lin, W.Y., Lee, H.C., and Palmbach, T.M., Forensic botany: using plant evidence to aid in forensic death investigation, *Croat. Med. J.*,46, 606–612, 2005.
5. Wawryk, J. and Odell, M., Fluorescent identification of biological and other stains on skin by the use of alternative light sources, *J. Clin. Forens. Med.*, 12, 296–301, 2005.
6. Budowle, B., Schutzer, S.E., Ascher, M.S., Atlas, R.M., Burans, J.P., Chakraborty, R., Dunn, J.J., Fraser, C.M., Franz, D.R., Leighton, T.J., Morse, S.A., Murch, R.S., Ravel, J., Rock, D.L., Slezak, T.R., Velsko, S.P., Walsh, A.C., and Walters, R.A., Toward a system of microbial forensics: from sample collection to interpretation of evidence, *Appl. Environ. Microbiol.*, 71, 2209–2213, 2005.
7. Li, R.C. and Harris, H.A., Using hydrophilic adhesive tape for collection of evidence for forensic DNA analysis, *J. Forens. Sci.*, 48, 1318–1321, 2003.

3 Techniques of DNA Fingerprinting

JOHN SCHIENMAN, PH.D.

Contents

3.1 Introduction to DNA Test Methods

The purpose of this chapter is to provide a basic understanding of the molecular protocols used for DNA fingerprinting or DNA profiling. There have been a number of techniques developed over the years, but this chapter will focus on the more current (or generally considered to be more consistently informative) techniques, namely the polymerase chain reaction (PCR), DNA sequencing, amplified fragment length polymorphism (AFLP), and microsatellite analysis of short tandem repeats (STRs). It will be assumed that the reader possesses a basic knowledge of the structure and chemical properties of DNA.

3.2 The Polymerase Chain Reaction

The polymerase chain reaction (PCR), first developed by Kary Mullis in 1986, is the basis of or a foundation component of the majority of techniques used in DNA fingerprinting.[1,2,3] For most molecular protocols involving DNA, an amplification of the DNA sequence to be analyzed is required. PCR is an amplification process that generates a sufficient copy number of the DNA region of interest, or target, allowing for the detection of a specific DNA sequence in a sample and for further analysis by such methods as DNA sequencing, AFLP, and STR. PCR is *in vitro* DNA replication and the essentials of the process are illustrated in Figure 1. However, understanding how PCR was developed requires a brief examination of *in vivo* DNA replication.

Eukaryotic *in vivo* replication of genomic DNA requires the appropriate ribonucleotides, deoxy-ribonucleotides, and the following enzymes: helicase, gyrase, RNA polymerase, and DNA polymerase. The nucleotides are the raw materials (or building blocks) that will be used in both the formation of short strands of RNA primers and then in the newly synthesized, longer strands of DNA. The helicase and gyrase function to unwind and separate (denature) the duplex strands of DNA that comprise each chromosome, the compressed "package" of DNA that is inherited. This allows the RNA polymerase to bind to these single DNA strands and synthesize short segments of complementary RNA. The result of this process is a hybrid duplex that consists of one strand of DNA and one strand of RNA with a free 3′-hydroxyl group. This short hybrid duplex with its free 3′-hydroxyl group is the target required by the DNA synthesizing enzyme, DNA polymerase. Once the DNA polymerase has found this target, it begins to move in a 5′ to 3′ direction, adding the appropriate deoxy-nucleotide to the growing chain complementary to the existing nucleotide of the opposite strand. This forms a double-stranded DNA molecule composed of one old strand and one new.

Simple, but ingenious, modifications to the *in vivo* process made it possible for DNA replication to be carried out, outside the physiological environment of a biological cell, in a plastic tube. One modification of the process is to use presynthesized DNA oligos to substitute as primers removing the need for RNA polymerase and ribonucleotides. Additionally, using temperature changes to denature the double-stranded DNA template and then to anneal the oligo-primers eliminates the need for the helicase and gyrase enzymes and, again, the RNA polymerase. Using temperature change to control the reaction requires one more important modification, which is the use of a thermo-stable DNA polymerase. A thermo-stable DNA polymerase, isolated from thermophilic bacteria, can withstand the temperatures necessary to denature double-stranded DNA (typically 94–95°C) and then retain the polymerase activity when the temperature is reduced. Additionally, since the optimal temperature for the activity of these polymerases is approximately 72°C, the reaction can be designed to amplify a very specific target using DNA oligos that will anneal in the temperature range of 50–65°C.

Typically, one is not interested in replicating an entire genome but only a very small portion of it, such as a single gene, segment of a gene, or some other small region of a genome. This interest in amplifying small specific segments brings forth another important feature of the use of short DNA oligos in place of RNA polymerase and ribonucleotides. Since DNA polymerases can

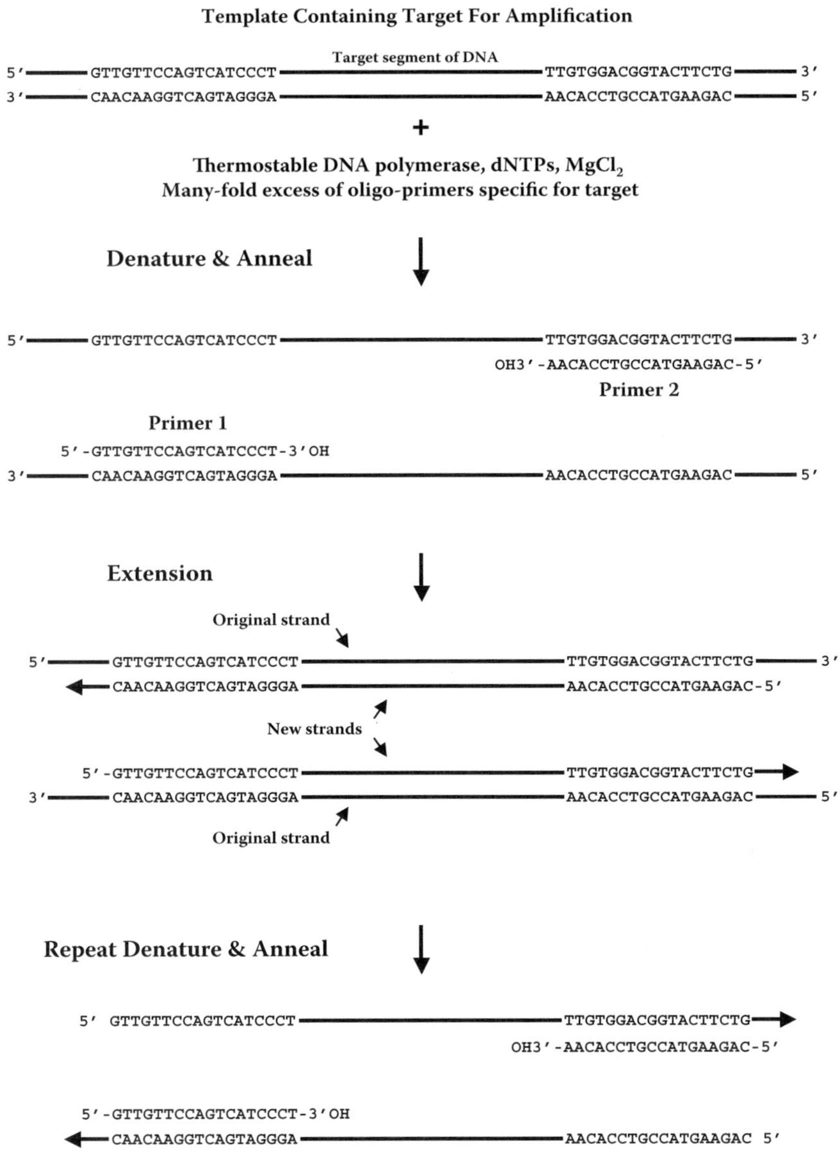

Figure 1 The process of PCR amplification of segments of DNA involves three main steps: 1) separation/denaturation of the DNA double helix at high temperatures (95°C), 2) annealing of short complementary DNA primer sequences that determine the specific region of DNA to be amplified. (50°–65°C), and 3) synthesis/extension (72°C), which completes the amplification process for a single cycle of PCR. Typically, forensic STR marker amplification involves 28–32 full cycles of PCR.

only synthesize a new strand of DNA in a 5′ to 3′ direction from a preexisting region of double-stranded DNA with a free 3′-hydroxyl group, the region of DNA that is replicated or amplified can be specifically targeted. This is accomplished by designing two short DNA oligos (typically 18–23 bases in length) that are complementary to regions of the genome bracketing the segment to be amplified. The only remaining requirement to being able to carry out this primer design is that enough DNA sequence information is known to construct the oligos.

In Figure 1, each black line represents a strand of DNA with the polarity denoted by the 5′ or 3′ designation at each strand end. The template can be any DNA source, genomic or cloned, that contains the target to be amplified. The DNA duplex at the top of Figure 1 represents a fragment of genomic DNA that contains the target. The only sequence of DNA shown is that to which the primers have been designed to hybridize or bind. One oligo (primer 1) is complementary to the bottom strand of the DNA duplex upstream of the target segment and the other (primer 2) to the top strand downstream of the target segment (Figure 1). After denaturing the double-stranded DNA template and annealing the oligo-primers to their complementary sequences, DNA polymerase will extend from each primer, creating a new strand of DNA that, if extended far enough, will contain the complementary sequence of the other oligo-primer. In this way, a newly synthesized strand extended from one of the primers can act as a template for hybridization of and extension from the other primer in the next cycle of the PCR reaction. Multiple repetitions, or cycles, of denaturing the DNA into single strands, annealing the oligo-primers to their complementary sequences, and extension with DNA polymerase creates a geometric progression of amplification, or doubling of the target DNA region in each cycle. In Figure 1, one can see that when the bottom original template strand is hybridized by primer 1, the newly created strand will always extend past the primer 2 binding site. But, when one of these newly synthesized strands is used as a template, the complementary strand extended from primer 2 will cease after the primer 1 binding site, since there is no template of DNA beyond this point for that strand. This new strand of DNA will extend only from primer 1 to primer 2. Similarly, when a top original template strand is annealed and extended with primer 2, the following cycle will create a new strand that extends from primer 2 to primer 1 only. In just a few short cycles of doubling the template strands, strands that extend just from one primer sequence region to the other will come to predominate the mixture of available template DNA strands, resulting in the amplification of a billion or more copies, or microgram quantities, of the target region of DNA between, and containing, primers 1 and 2 after 30–40 repeats of the temperature cycling.

In practice, one designs the oligo-primers using one of the many software programs designed for this task. Despite the years of work optimizing the algorithms of these programs to produce the best possible primer pairs for DNA amplification, one sometimes still needs to empirically optimize the PCR reaction for a particular set of primers and template. The two most basic parameters that influence the specificity and efficiency of amplification of the target DNA sequence include annealing temperature and magnesium chloride concentration. Both of these parameters influence the hybridization kinetics of primers binding to the template DNA. Magnesium chloride actually has two functions in the PCR reaction. First, it is a cofactor for the DNA polymerase and, therefore, is required for the enzyme to function. Second, the positively charged magnesium ions will electrostatically shield the negatively charged phosphates of the sugar-phosphate backbone of each DNA strand. Hydrogen bonding between the nitrogenous bases and base stacking are the forces holding two complementary DNA strands together. But, the negatively charged phosphates of the backbone create a slightly repulsive force between the two strands of DNA. The shielding by magnesium will reduce this repulsive force making the double-stranded structure of any DNA more stable. Increasing temperature makes double-stranded DNA less stable while increasing magnesium concentration has the opposite effect. One optimizes the concentration of magnesium and the annealing temperature so the PCR reaction amplifies the target DNA segment and no others from the other regions of the chromosomes. The magnesium concentration being too high and/or the annealing temperature being too low increases the probability of either primer annealing to a close, but not exact, complementary sequence match. If this occurs on opposite stands close enough for amplification, then segments of DNA other than the target can be amplified. This will likely reduce the amount of desired target that gets amplified as well as generate a mixture of PCR products that all have the primer sequence/s at their ends. This type of product mixture will not be useful for further analysis by other methods, as those methods are based on the condition of only intended amplification products being present.

Optimizing the magnesium and annealing temperature conditions usually results in yield of the target DNA that is useful for other applications, and one might assume that this optimization produces a PCR reaction that is 100% efficient. As discussed previously, if PCR is 100% efficient, it results in a doubling of target DNA in each round of amplification. But, typically, PCR for any primer set is rarely 100% efficient. A reaction would not be 100% efficient if all available target DNA strands are not hybridized by the primers in each round of amplification. For example, if on average 75% of the target strands are annealed by a primer and 25% are not, then instead of a doubling, the target will increase by 1.75 times in each cycle. Some of the reasons for less than 100% primer annealing are secondary structure (e.g., folding of the DNA) of the DNA template strand inhibiting primer annealing or secondary or hybrid structures of the primers themselves (e.g., primers binding to primers). For most applications, this is not a concern because 30–40 cycles of amplification are still sufficient to make large quantities of target despite the somewhat reduced efficiencies of the reaction. Different efficiencies of different primer and/or allelic target sets explain why quantities of the PCR amplicons produced by these sets are typically not equal when performed in a multiplexed reaction (i.e., where primer sets are mixed together to attempt to amplify more than one target sequence in the same PCR reaction) in which the initial target copy number of each locus or allele (segment of DNA) is identical.

3.3 DNA Sequencing

The current technique of DNA sequencing is a variation on the PCR theme, namely the PCR primer extension reaction, with three main differences.[4,5] The first, is only one oligo-primer is used in the reaction, so primer extension occurs for only one of the two strands of the double-stranded DNA template. This means that, although we are making DNA during the sequencing process, we are not amplifying it in a geometric fashion. For this reason, the second main difference is that a much larger starting quantity of template is required. For PCR, the starting copy number of target DNA molecules should be greater than 1,000 to yield microgram quantities of the final product. For sequencing, approximately 5×10^{10} copies of target molecules are needed to generate good quality fluorescent signals to "read" the target DNA. So, typically, a PCR reaction is performed first to generate enough templates for direct sequencing, or for cloning followed by sequencing of the clone. The third main difference is the addition of fluorescently labeled dideoxy-ribonucleotides along with standard deoxy-ribonucleotides. These labeled dideoxy-ribonucleotides serve two functions. In chemical nomenclature, dideoxy means that these nucleotides have a hydrogen atom attached to the 3′-carbon of the sugar instead of the hydroxyl group of the standard deoxy-nucleotide. Since the DNA polymerase enzyme can only extend a DNA strand from a free 3′ hydroxyl group, any DNA chain having a dideoxy-nucleotide incorporated into it will terminate at that nucleotide base. So, its first function in the reaction is as a DNA chain terminator. The second function involves the fluorescent label, or tag. This tag serves to produce a detectable signal when excited by the appropriate energy source, such as a laser. Since DNA is invisible to the naked eye, fluorescent tags allow for the detection of each nucleotide base.

Figure 2 illustrates the basic process of DNA sequencing. We begin with a double-stranded DNA template, for example, the PCR product generated

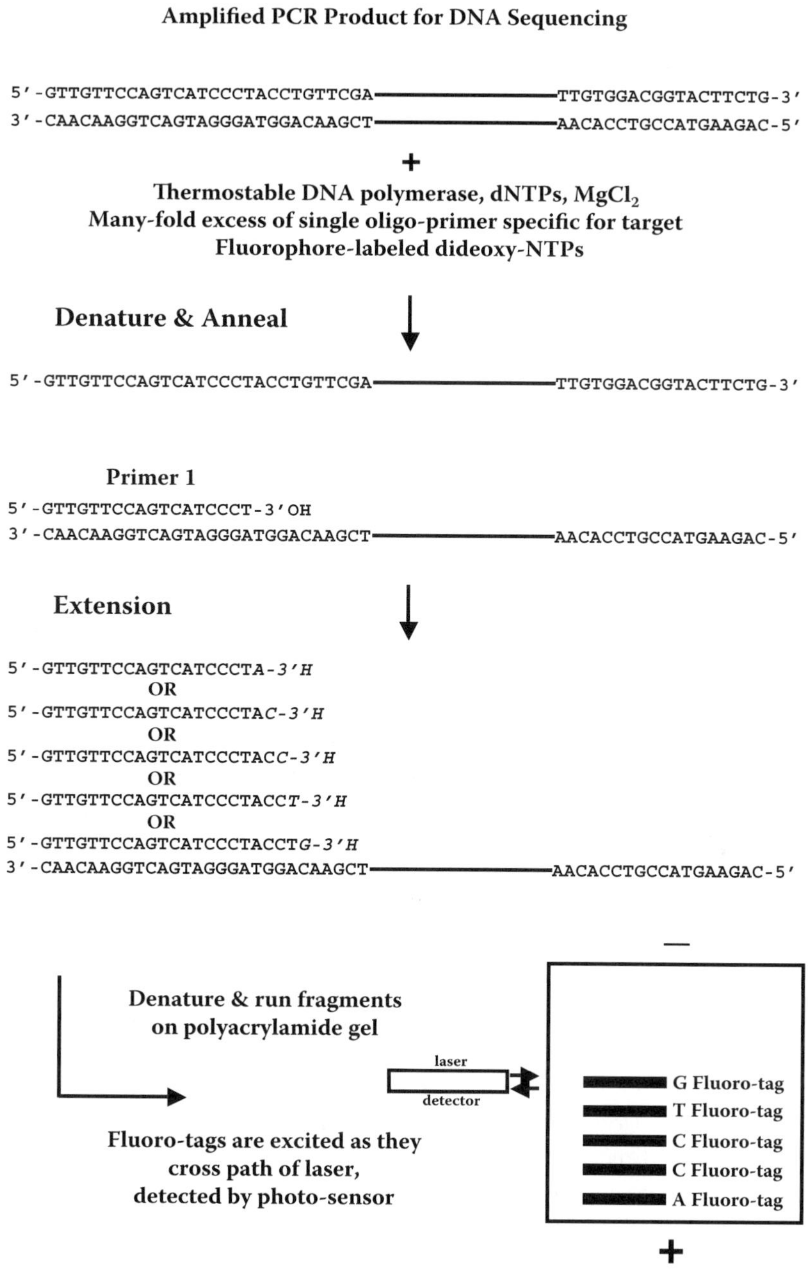

Figure 2 During the PCR extension step of DNA sequencing, a fluorescent tag is added with the incorporation of a chain terminating dideoxy-NTP into the extending DNA strand. This detection step allows for reading the order of nucleotide bases in a DNA fragment. Testing to detect single-base differences to determine whether an organism could be the source of the DNA is an important feature of mitochondrial DNA sequencing.

in Figure 1. For ease of explanation, only the subsequent 10 nucleotide bases beyond the primer 1 binding site are shown. The DNA template is denatured by heating followed by a lower temperature of 50–55°C to anneal primer 1 to the bottom strand of the template. This primer, as before, has a 3′ hydroxyl group so DNA polymerase will begin to add nucleotides complementary to this bottom strand. The deoxy/dideoxy-nucleotide ratio added to the reaction is such that the probability favors the additions of deoxy-nucleotides with the occasional incorporation of a dideoxy-nucleotide. As soon as this second type of nucleotide is added to the growing chain, the chain is terminated. Again, for clarity of explanation, the extension section of Figure 2 shows only one fragment for the first five possible termination products, in order from smallest to largest. However, this dideoxy-nucleotide incorporation is essentially random and many, many fragments of each possible termination product will be produced in the reaction over the course of 25 temperature cycles of denaturing, annealing, and extension.

The first product shown in Figure 2 is terminated at the first base addition following the primer strand, an A, with its total length in nucleotide bases now equaling 19. The second is terminated at the second base addition, a C, with a length of 20, and so on. This fragment mixture is then denatured one last time and loaded onto a vertical polyacrylamide gel. The fragments migrate down through the gel under the force of an electric current. Since DNA strands have an overall negative charge due to the phosphates in the backbone of the molecule, they will run towards the positive pole. Intuitively, one might first surmise that longer fragments of DNA run faster throw the gel matrix because they carry more negative charges. But, in fact, the opposite is true; shorter fragments run faster. This is because the charge has no effect on the rate of migration for different-size molecules, because every DNA molecule always has essentially the same charge-to-mass ratio, that is, every base has one phosphate. Obviously, adenine (A), guanine (G), cytosine (C), and thymine (T) do not all have the same mass, but unless a strand is made up from predominantly one nucleotide relative to another strand, the mass difference is insignificant. So, two fragments of the same length will run at the same rate and the longer the fragment, the slower its migration through the gel. The density of polyacrylamide, and thus the size of the pores or holes through which the DNA strands move, is measured as a percentage. A standard gel for DNA sequencing is composed of 5–6% polyacrylamide and this percentage is capable of resolving DNA strands that differ in length by a single nucleotide base.

The last part of Figure 2 shows an illustration representing a sequencing gel with the DNA fragments generated from our hypothetical PCR reaction product. The shortest fragment has a single dideoxy-A nucleotide added to the end of the primer chain, and thus these fragments would be the first to migrate past the fixed vertical position of the laser energy source and photo-detector.

The second set of fragments that would migrate past the laser/detector would be the ones that had a deoxy-A and then a dideoxy-C added, and so on. Since the four dideoxy-nucleotides each have a unique fluorophore (fluorescent dye tag), which when excited by the laser energy source will each emit a slightly different wavelength of light, the wavelength of light detected determines what dideoxy-nucleotide was at the 3′-end of that set of fragments. The wavelength, amplitude, and duration of light being detected are stored in a computer file for analysis at the end of the gel run. So the DNA sequence in the example of Figure 2 would be read A, C, C, T, and G, etc.

Examples of the graphical output from analysis of such a computer sequence file, or electropherogram, can be seen in Figure 3. Two of the

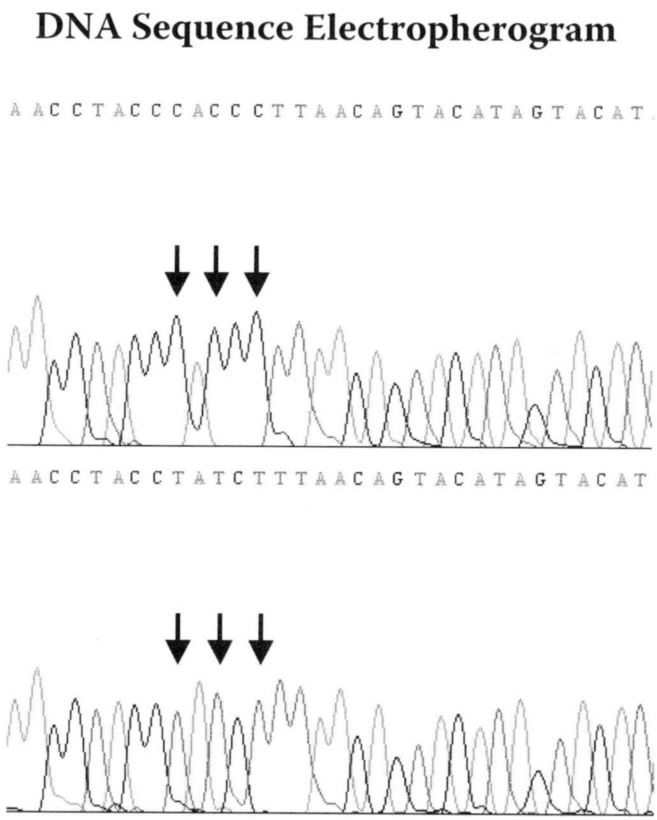

Figure 3 An example of DNA sequence data is shown here. Each nucleotide base (C, G, T, A) is assigned a specific color (e.g., C is blue) for ease in interpretation of the sequence by the DNA alignment software.

most common DNA regions sequenced for genotyping purposes are the hypervariable regions I and II of the mitochondrial genome of eukaryotes.[6] Figure 3 shows portions of electropherograms (data courtesy of Josh Suhl, University of Connecticut) of a DNA sequence from a 32 base segment of hypervariable region I for two unknown human individuals. Each peak represents the amplitude and duration of one of the four frequencies of light detected during the sequencing run. The four different colored peaks represent the four different wavelengths produced by the flourophores associated with each dideoxy-nucleotide. The sequence of the individual depicted in the top panel has a C base in the human mitochondrial sequence positions 16,292, 16,294, and 16,296, respectively.[6] In contrast, the sequence of the individual depicted in the bottom panel has a T base at these three positions.

3.4 Amplified Fragment Length Polymorphism

Amplified Fragment Length Polymorphism (AFLP) is an extremely useful method for genotyping individuals of species where little or no genome sequence data is available. Unlike the RAPD method, which directly generates from PCR a number of different length DNA fragments from an individual using six-base long (hexamer) primers, AFLP first creates these fragments by enzyme digestion at specific DNA sequence sites.[7] Because this type of fragment generation is not dependent on any of the factors that can influence PCR efficiency, the method is less sensitive to slightly variable reaction conditions and thus more reproducible.[8,9] In AFLP, the genome is first treated with specific DNA restriction enzymes, which will cut it into a consistent set of fragments. Then DNA linkers (a short specific sequence of double-stranded DNA) are ligated or added onto the ends of these fragments. With the attached linkers, the fragments will all have the same two 20–30 base pairs of DNA sequence at their ends and can now be amplified with just two specific oligo-primers.

This protocol relies on fragment generation by DNA restriction enzymes, which, if performed appropriately, can generate a set of fragments unique to an individual's genome. DNA restriction enzymes are isolated from prokaryotes where they are thought to have evolved as a primitive immune system to destroy invading foreign DNA, such as that from bacteriophages. The host organism protects its own DNA from cleavage by these enzymes by nucleotide modifications, such as methylation, within its own genome. The most commonly used DNA restriction enzymes are type II endonucleases. These nucleases cut at internal sites within a piece of double-stranded DNA and typically cut at a very specific sequence of nucleotides, or "recognition sequence". This recognition sequence, in which the enzyme will cut the sugar-phosphate linkage of both strands, is variable in length depending on the enzyme but, four to six base recognition sequences are most common. The number and size of fragments generated by a particular enzyme cutting a larger piece of DNA are dependent on the complete DNA sequence itself. As long as this DNA sequence remains unchanged, so will the pattern of digestion. For this reason, restriction enzymes were one of the first diagnostic tools developed to characterize and identify (i.e., "fingerprint") specific pieces of DNA.

Although there are hundreds of DNA restriction enzymes commercially available, the two most commonly used enzymes for AFLP are EcoRI and MseI. The naming of these enzymes is based on the name of the organism from which they were originally isolated, (e.g., EcoRI was isolated from *Escherichia coli*). EcoRI recognizes the six-base sequence 5′-GAATTC-3′ and MseI recognizes the four-base sequence 5′-TTAA-3′. Note that these recognition sequences are often palindromic (i.e., the sequence reads the same when it is read in a 5′–3′ direction on either DNA strand). Based solely on the probability that one would expect to find these sites in a random sequence of DNA, the average base pair distance between two of the same recognition sequence can be calculated. Given there are only four possible nucleotide bases, the probability of finding a particular base at a particular nucleotide position of a DNA sequence is $^1/_4$. The probability of a specific sequence of more than one base is simply determined by multiplying the probability of each individual base in the sequence. So a specific four-base sequence, or recognition site, should occur on average once in every $(4)^4$ or 256 bases. Likewise, a specific six-base sequence should occur once in every 4,096 bases. Considering the fragment resolution capability of acrylamide gel electrophoresis and the goal of producing some fragments unique to an individual of a species, the size range of fragments that are useful for genotyping using AFLP is approximately 50–500 bases in length. Using one six-base recognition site restriction enzyme and one four-base recognition site restriction enzyme will generate a large number of fragments in this size range.

Along with recognizing a short specific DNA sequence, another feature of DNA restriction enzyme cleavage of double-stranded DNA is the site (i.e., between which two bases) where the sugar/phosphate backbone linkage is broken. Many of the enzymes cut the backbone in a symmetric, but staggered fashion, producing a cut with an overhang of one strand (see Figure 4). An advantage of this feature is that fragments that have been cleaved can also be ligated or "glued" back together as long as they have compatible complementary overhangs. These staggered cut DNA ends are frequently called "sticky end overhangs" in molecular biology jargon.

Figure 4 diagrams the basic process of the AFLP method. Genomic DNA is first digested with two enzymes, one a six-base restriction enzyme (EcoRI) and the other a four-base restriction enzyme (MseI). A typical DNA isolation

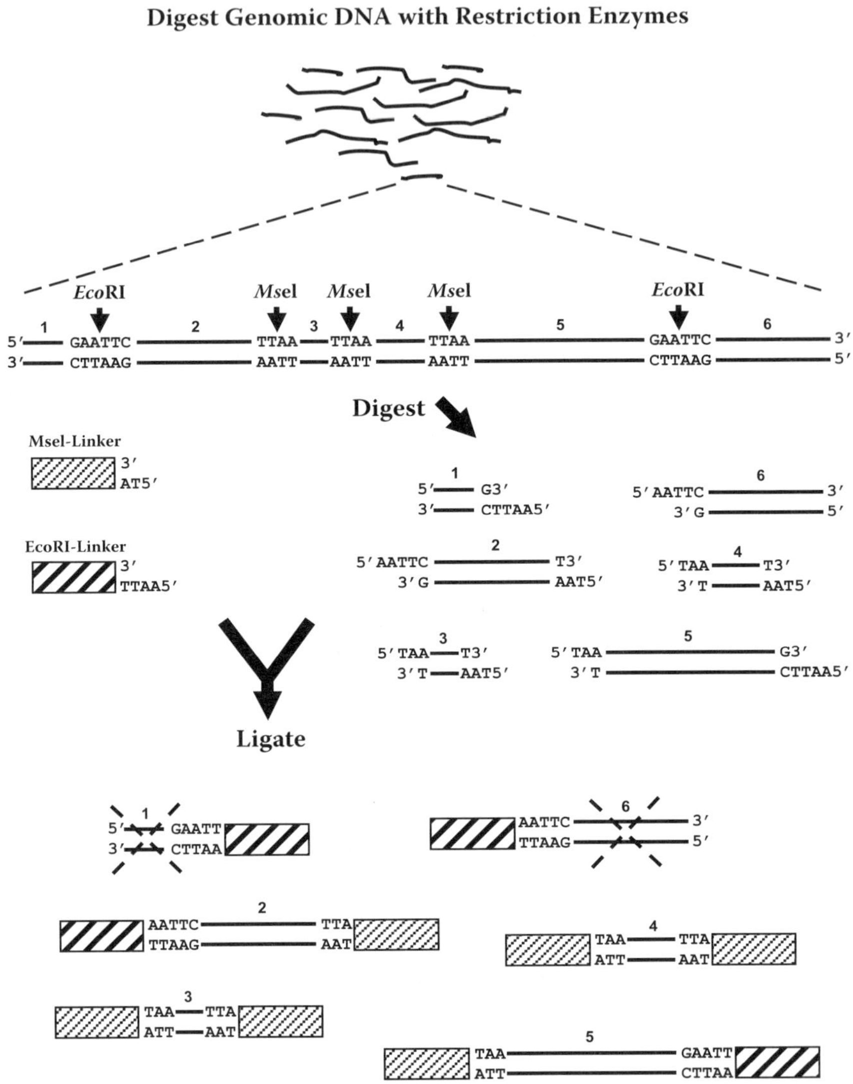

Figure 4 AFLP analysis is a DNA typing technique that will generate a DNA fragment profile from almost any organism of interest. DNA fragments are generated by restriction enzyme digestion, adaptor sequences are ligated on the matching ends of the fragments, two rounds of PCR amplify the fragment population, and finally, a subset of fragments are detected during capillary electrophoresis.

protocol is not going to isolate whole intact chromosomes, but randomly sheared chromosome fragments between 20,000–100,000 base pairs in length. The irregular lines at the top of the figure represent these large pieces of DNA that will be cleaved into many much smaller pieces. DNA linkers and ligase

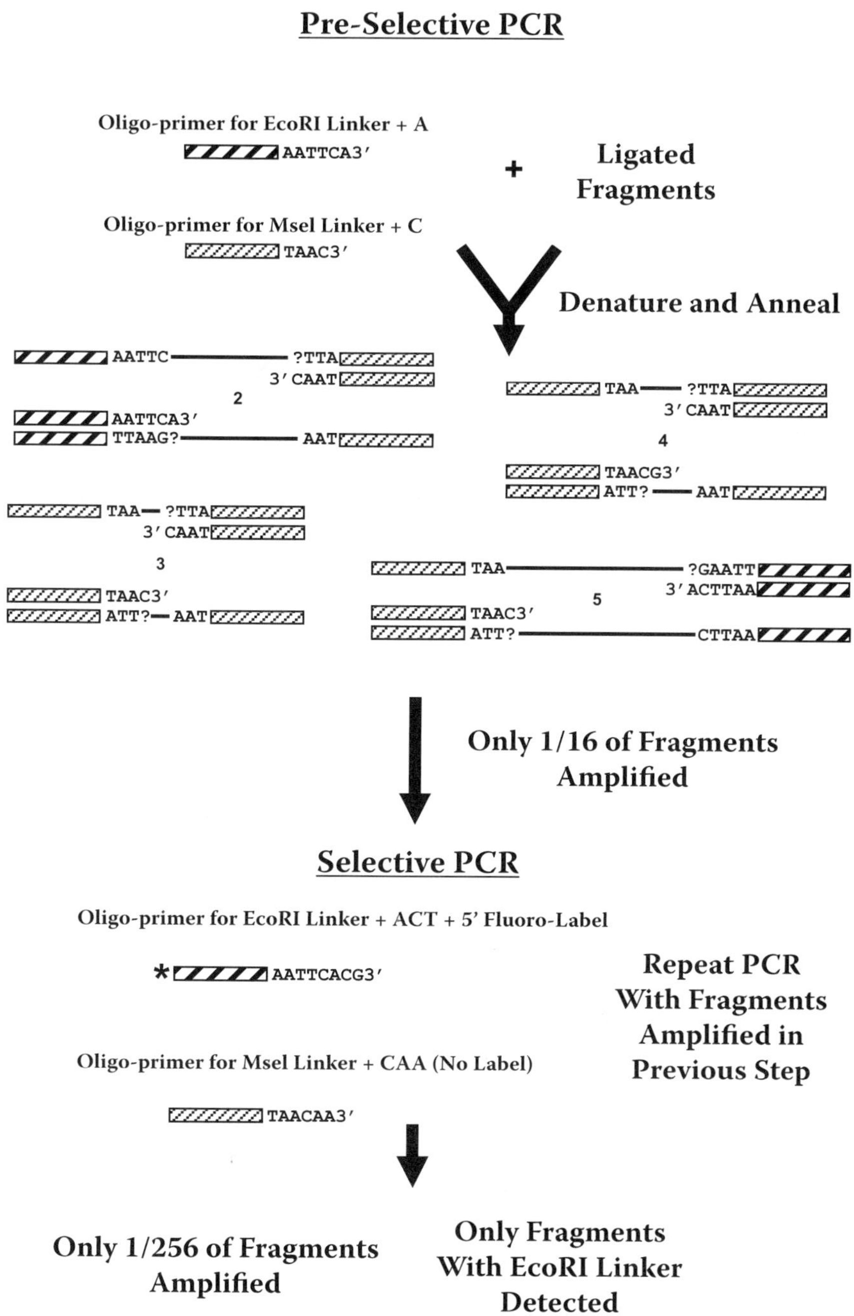

Figure 4 (Continued)

are also added to the enzyme cleavage step. Compatible complementary overhangs allow the linkers to be ligated to the cleaved genomic fragments. The linker sequence is designed such that the last base in the linker before the overhang does not match the consensus base for the restriction enzyme recognition site. The ligation of linker to the genomic fragment results in the loss of the restriction enzyme's recognition sequence. In contrast, if two EcoRI digested genomic fragments are ligated, the site is not lost and can be recleaved by the restriction enzyme. This allows for the digestion of genomic DNA and ligation of linkers to be carried out at the same time and eliminates the possibility of concatenated (i.e., tandemly "glued") genomic fragments. Upon digestion, fragments that are produced from the ends of the initial large genomic fragments (fragments such as 1 and 6 in Figure 4) will not have a linker ligated to both ends, because the one end was not produced by enzyme cleavage, but by shearing during the DNA isolation procedure. Thus, this type of fragment can never be PCR amplified with linker-specific primers.

At this point there are too many fragments to separate and analyze on an acrylamide gel. There are literally millions of different fragments produced for each copy of a billion-base-pair-long genome treated in this way, with several representatives of each possible fragment length from approximately 10,000 base pairs on down in the size range. A large reduction in the number of fragments is necessary for a meaningful analysis. This is accomplished by two sequential PCR reactions with extra bases added onto the 3′ ends of the primers. In the first preselective PCR reaction, one additional base is added to the forward and reverse primers. The random probability of a matching complement base in the next base pair position downstream of the primer is $^1/_4$. The same is true for both the forward and reverse primers, so you simply multiply to get the combined probability, or $^1/_{16}$. Preselective PCR results in amplification of approximately $^1/_{16}$th of the fragments produced by the restriction enzyme digestion. The '?' in the DNA strands of Figure 4 represents the condition that each primer will only be extended into a new DNA strand if that base is a complementary match to the 3′-base of the primer. These amplified fragments are then used as a template in a second selective round of PCR, in which, in addition to the previous extra base added to each primer, two more bases are added to the 3′-ends. This accomplishes an additional reduction, such that only $^1/_{256}$ of the preselectively amplified fragments will be reamplified. The overall reduction in number of the DNA fragments produced by enzyme digestion is approximately $^1/_{4096}$ and might, typically, amplify somewhere between 20 and 30 fragments in the 50–200 base pair range. The goal of these two amplification steps, beyond the amplification itself, is to reduce the number of fragments so that electropherogram analysis of the acrylamide gel is manageable, but still allowing for detection of differentially present fragments unique to an individual genotype. A key detail of the process is that the selective EcoRI primer is labeled with a flourophore so it can be detected by a laser/photo-sensor system. Fragments with MseI linkers on both ends will be amplified but will never be visualized in the final analysis, since there is no flourophore label associated with the MseI primer. Fragments with EcoRI linkers on both ends of the DNA fragment are possible, but unlikely given the expected frequency of MseI recognition sites in any given DNA sample. Before loading the individual samples on an acrylamide gel, each sample gets mixed with prelabeled size standards with fragments ranging in size from 50 bases up to 500 bases. This allows for the analysis software to account for slight differences in gel running conditions from lane to lane and thus properly align the samples for lane-to-lane comparison. Despite the fact that no prior genome sequence information is needed to use AFLP as a genotyping method, there is some initial optimization required for each species it is applied to. The goal of the fragment generation is two-fold: 1) production of a low enough number of different-size fragments such that they can be easily resolved and analyzed (i.e., too many bands, especially compressed bands, are difficult to interpret); 2) production of enough different-size fragments such that fragments unique to an individual are generated (i.e., a sufficient number of markers are required to individualize a sample). Because the genome of each species is unique, there is no guarantee that any given forward and reverse primer set will generate a useful set of fragments for every species. There are 256 possible forward and reverse selective primer combinations if just the second and third base additions are considered. Typically, eight different forward and eight different reverse selective primers are supplied with kits. In practice, several combinations would be tried on a few individual samples to determine what combinations will be useful for larger scale analysis.

Figure 5 provides an example of an AFLP profile generated from one set of selective PCR primers for two marijuana plant samples. As in DNA sequence analysis, peaks represent DNA fragments of various lengths that have an attached flourophore. In contrast to sequence analysis, the flourophore tag is incorporated into the DNA chain as part of one of the selectively amplifying primers. Additionally, there is not a ladder of fragments differing in length by a single nucleotide as in DNA sequence analysis, but rather a random distribution of various lengths based on the distribution of endonuclease recognition sites (thus the DNA sequence) of the genome of each sample. For this example, peaks are shown for a range of 69–182 nucleotide bases. The relative fluorescence unit (RFU) levels (seen on the Y-axis to the right of each sample) of each peak are proportional to the amount of amplification of that fragment during the PCR reaction. Peaks are only considered for genotyping analysis if they have a height above some user-defined fluorescence level. A typical cutoff level might be 50 relative fluorescence units.

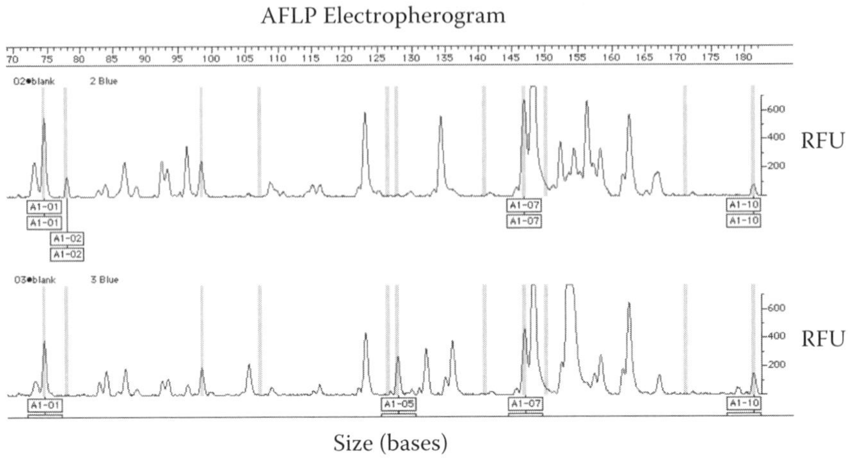

Figure 5 A section of an AFLP electropherogram that shows DNA fragments in the size range of 70–180 nucleotide bases that have been tagged with a blue fluorescent label for visualization. The Y-axis is expressed as RFU, relative fluorescence units, to indicate the intensity of the fluorescence of the DNA fragment.

Although most of the amplicons were generated by the same primer pair, length and sequence differences between each amplified genomic fragment can result in different amplifying efficiencies as previously discussed.

Ultimately, a direct comparison by eye of an alignment or overlay of the fragments would be used to determine if two samples were consistent with originating from the same or different genomic DNA sources. But, the fragment data will often be overlaid with specific-size bins for database storage and faster, automated, computational database searching and retrieval. In this particular example, 10 bins have been predefined, based on previously generated data to determine fragments whose amplification with this selective primer set is polymorphic for this species, i.e., in some individuals this genomic fragment is generated by endonuclease cleavage and is thus amplified, while in others it is not. The top sample has an amplified fragment present for bins 1, 2, 7, and 10, while the bottom sample is positive for bins 1, 5, 7, and 10, establishing that these two samples did not come from the same individual, or clonally derived, plant.

3.5 Short Tandem Repeats

The tandemly repeated DNA units of mini- and microsatellite loci are often very useful for genotyping due to their typically high level of polymorphic variation in a population. This last section will discuss what these loci are

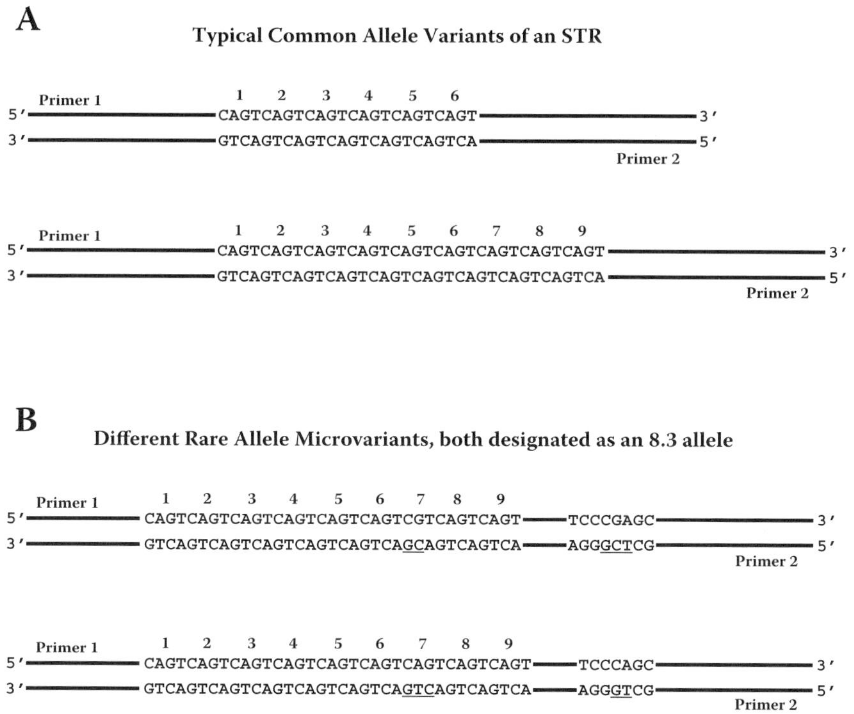

Figure 6 A) An illustration of short tandem repeat (STR) markers. The top panel indicates six short tandem repeats; the bottom panel has nine repeat sequences. B) Occasionally, variations in DNA sequences occur such that a full four-base repeat difference is not observed. In those cases, incomplete repeat sequences are reported as the number of full repeats plus the number of extra bases (e.g., 8.3 = eight full four-base repeats and three additional bases).

and what must be taken into consideration to properly amplify and interpret the results when using them for genotyping. Microsatellite sequences, now more commonly referred to as short tandem repeats (STRs), have a repetitive unit of two to six bases in length, repeated in a tandem or head-to-tail orientation (see Figure 6). The satellite nomenclature comes from early studies in which genomic DNA was isolated and then fractionated using density gradients. Fractions were analyzed with spectrophotometry and then each fraction's density was plotted against their absorbency values. It was found that the bulk of the genomic DNA was collected in one fraction and produced the main absorbance peak, but there were also one or more secondary, or satellite, absorbance peaks. These fractions were found to contain AT-rich repetitive DNA sequences typically associated with the centromere or telomere regions of chromosomes. Satellite DNA soon came to mean any tandemly repeated DNA. The mini and micro prefixes were used for repetitive DNAs that were composed of shorter repeat units with a lower copy number of this unit. Figure 6A provides an example of two possible alleles of a hypothetical locus with a CAGT repeat. One allele has six copies of the repeat sequence while the other has nine copies. Regions of conserved sequence just upstream and downstream of the STR locus are used to design primers for PCR amplification of that site. Because any two alleles will typically differ only in the copy number of the STR, the difference in length of each amplicon will be whole multiples of the four-base repeat. For this reason, alleles are designated by the number of tandem repeats they contain. The reason these loci are typically so polymorphic and the alleles often differ in length by whole multiples of the repeat unit is due to strand slippage or stutter during replication.

Experimental evidence supports the idea that when an extending DNA polymerase is released prematurely, the incomplete DNA strand can denature from the template and then reanneal.[10] If this occurs in the region of the repeated units, the extended strand can anneal in a displaced, or out-of-register, fashion due to the repetitive nature of the sequence. The fact that the base complementarity is a short unit in a tandem organization, a small kink in either strand allows for annealing of the last few bases of the new strand to a repeat unit preceding or following the one it was first replicated from. If a new polymerase molecule begins extending from this displaced strand, a DNA duplex with strands of unequal length will be generated with the length difference being some multiple of the repeated unit. A size difference of a single repeat unit is the most common. *In vivo*, these unequal strands will be corrected by DNA repair mechanisms usually back to the length of the original allele. Occasionally, the DNA duplex can be repaired such that a new allele is generated. If this occurs during the formation of a germ cell and this germ cell becomes part of a zygote, a new allele or mutation is generated. During *in vitro* DNA replication (PCR), these unequal strands will be denatured and used as templates in the next round of amplification and thus will result in a mixture of PCR products. When analyzed on an acrylamide gel, there will not only be a peak representing the true size of the allele, but one or more peaks representing PCR stutter products that differ in size from the true peak by whole multiples of the repeat unit. Experimental evidence has shown that longer repeat units are less susceptible to the production of stutter amplicons.[10] This is why STR loci of four bases or more are used for forensic applications.[11] Stutter can still occur for such loci (see Figure 7), but the amplification of such products is typically fewer than 10% of the true allele as measured by peak height. Levels of stutter higher than this would be a significant problem when trying to determine if a sample of genomic DNA from an unknown source was from a single individual or a mixture of different individuals (not an uncommon occurrence at a crime scene). If two individuals were contributors to a DNA sample and the contribution of one individual was small in comparison to the other, then the

STR Electropherogram

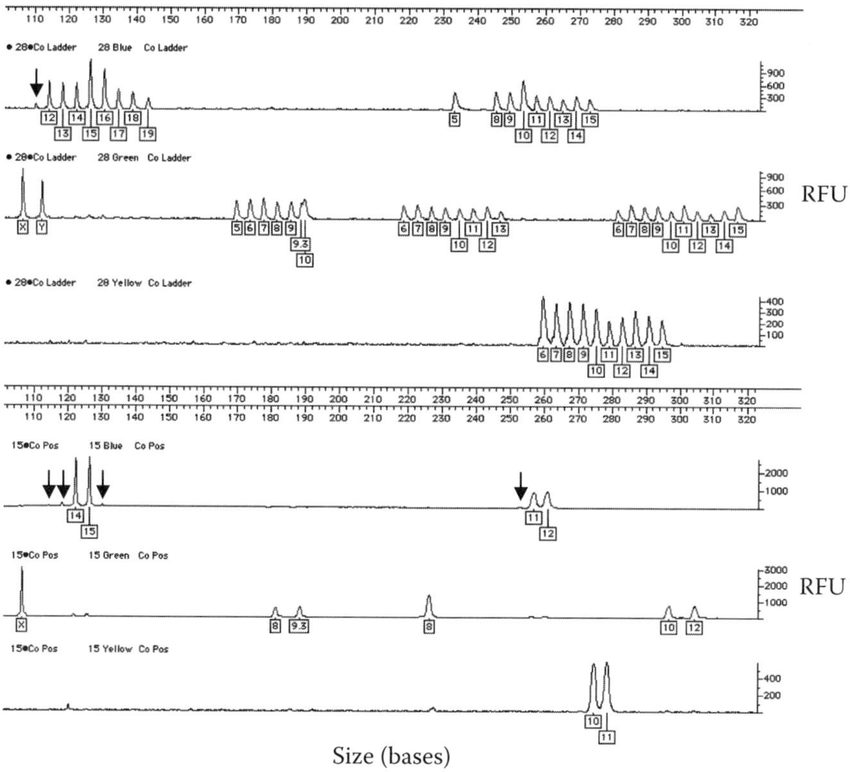

Figure 7 An example of an STR electropherogram with a commercially available allelic ladder (mixture of DNA fragments of known size for comparison to test samples) and a positive amplification control used to confirm that the STR kit is performing as expected.

smaller peak heights of the allele amplicons of the minor contributor could be mistaken for stutter products or vice versa. Thus far, we have discussed lengths of STR alleles always being whole multiples of the repeated unit due to a mutational mechanism caused by stutter. Obviously, other types of mutational events occur in genomic DNA and can thus occur in an STR allele. Two such events are point mutations and insertion or deletions of nucleotide bases. A point mutation is when a base pair is changed from one form to another; for example, an A-T base pair mutating to a G-C. Insertion or deletion mutations are exactly that, and can be of one or more base pairs. The impetus for such mutations can be exposure to mutagenic agents or spontaneous due to chemical tautomeric shifts of the nucleotides during replication. If such mutations occur in an STR allele, then there will either not be a size change or the size change will most likely not be a whole multiple of the repeated unit. Such STR allelic variants are known as microvariants. Figure 6B illustrates two possibilities for a deletion variant. The deletion is of a single base pair in both examples. In the first example, the deletion is an A-T base pair from the seventh CAGT repeat of the original allele, while the second is a C-G base pair in the region outside of the repeated units, but still within the region amplified by the primers. Amplicons of both of these alleles would be the same length and would be known as 8.3 alleles since they are one base pair shorter than a 9 allele. The only way to determine that these 8.3 alleles are actually different would be to sequence them. A number of such allele types have been recorded for many STR loci in use today.[12] Microvariants, while noted, do not interfere with the ability to type an individual and in fact often lend an extra bit of uniqueness to a DNA STR profile. Another type of amplicon, or peak, artifact that can occur in STR analysis is known as nontemplate addition. The most commonly used thermostable DNA polymerases have the propensity to add an extra A base to the 3′ strand ends of the PCR amplicons. When this occurs, the denatured amplicon strands will obviously be one base longer than the length spanned from one primer to the other. This is not a problem as long as this occurs to all of the amplicon molecules. This would make every molecule one base longer and thus there would be no relative change from one fragment to another. If both types are present in the amplification, then a double peak, or a peak with a shoulder, will be produced in the gel separation and analysis. Because the frequency with which this occurs can vary due to the amplification conditions, amplification protocols are designed to produce 100% nontemplate addition so only a single amplicon size, or peak, is produced for each allele. This is accomplished by putting enough nucleotides into the reaction so they are not a limiting factor, using the appropriate amount of genomic DNA template, and by adding a final 60°C or 72°C extension step of 30–45 minutes in duration at the end of the amplification temperature cycling profile. This ensures that almost every amplicon molecule has an A base added to both its 3′ strand ends.

Figure 7 is an electropherogram (i.e., software output) for a human commercial STR kit, COfiler (Applied Biosystems, data courtesy of Craig O'Connor, University of Connecticut). The kit contains reagents to amplify six STR loci (D3S1358, D16S539, THO1, TPOX, CSF1P0, and D7S820) and one sex chromosome locus (Amelogenin) from human genomic DNA. Amelogenin is not an STR but allows for sex determination of the contributor of an unknown genomic sample. The gene exists on both the X and Y human chromosomes, but the X version has a six-base-pair deletion relative to the Y version, allowing for size separation during electrophoresis if both are present. Just as for AFLP, when separating STR amplicons, an internal lane size standard (GeneScan 500ROX, Applied Biosystems Inc.) is added to each sample lane to allow for adjustment of slight lane-to-lane differences during electrophoresis. In addition, since virtually all the alleles present in human populations for these loci are known, an allelic ladder is loaded into several lanes (along with the same internal lane size standard). Using different flouro-phore tags for loci that have some allelic amplicons within the same size range allows for more loci to be amplified in a single reaction tube and analyzed in one lane of the gel. A direct comparison between lanes containing the allelic ladder and those containing an unknown sample generates a DNA profile of the individual for these seven loci. For the human sample shown in Figure 7, the individual would be typed: female (lack of Y-allele sized amplicon); (14,15) D3S1358 heterozygote; (11,12) D16S539 heterozygote; (8,9.3) THO1 heterozygote; (8,8) TPOX homozygote; (10,12) CSF1P0 heterozygote; and (10,11) D7S820 heterozygote.

If enough previous data of genotypes of many individuals from many populations have been collected, estimated allele frequencies within the populations can be calculated. Using these estimated allele frequencies, an expected genotypic frequency can be calculated for each locus. Where p and q represent allele frequencies, p^2 or 2pq (homozygous or heterozygous conditions, respectively) would be used to calculate the expected frequency of that particular genotype for each locus. To generate an expected frequency for all seven loci combined, one would take the product of the expected genotypic frequencies for each individual locus. The main impetus for making such a calculation in forensics is for the benefit of a typical layperson that would be sitting on a jury. Any DNA expert would recognize the full significance of a suspect sharing the same DNA profile as that left at a crime scene, and that the probability of two individuals (except for identical twins) matching at all seven loci is essentially zero. But obviously, given that it is a probability estimate, it is still within the realm of possibility. In fact, most forensic laboratories report STR profiles for a standardized set of 13 loci. Therefore, to be able to communicate the significance of a suspect being included as a donor of a DNA sample, the expected frequency of that genotype in the human population is calculated and reported as a random match probability.

3.6 Summary

Although many different DNA fingerprinting systems are available, the ones discussed in this chapter are those most commonly used in the forensic individualization of biological evidence, both from human and nonhuman sources. While STR marker systems are uniformly utilized to identify human DNA left at crime scenes, they are also becoming more common for nonhuman DNA sources such as selected plant species, cats, and dogs. For organisms that do not have developed STR systems, AFLP technology is a good alternative for any single-source, high-quality DNA sample. As the technology and court acceptance of nonhuman evidence progresses, more and more often will these forms of evidence be useful and presented for forensic casework resolution.

References

1. Mullis, K., Faloona, F., Scharf, S., Saiki, R., Horn, G., and Erlich, H., Specific enzymatic amplification of DNA in vitro: the polymerase chain reaction, Cold Spring Harbor Symposium in Quantitative Biology, 51(Pt 1), 263–273, 1986.
2. Saiki, R.K., Scarf, S., Faloona, F., Mullis, K.B., Horn, G.T., Erlich, H.A., and Arnheim, N., Enzymatic amplification of beta-globin genomic sequences and restriction site analysis for diagnosis of sickle-cell anemia, *Science*, 230, 1350–1354, 1985.
3. Mullis, K.B., The unusual origin of the polymerase chain reaction, *Sci. Am.*, 262, 56–61, 64–65, 1990.
4. Sanger, F. and Coulson, A.R., A rapid method for determining sequences in DNA by primed synthesis with DNA polymerase, *J. Mol. Biol.*, 94, 441–448, 1975.
5. Dideoxy Sequencing of DNA, http://whfreeman.com/biochem5/cat_040/ch06/ch06xd02.htm.
6. Brandon, M.C., Lott, M.T., Nguyen, K.C., Spolim, S., Navathe, S.B., Baldi, P., and Wallace, D.C., MITOMAP: a human mitochondrial genome database—2004 update. *Nucl. Acids Res.*, 33(Database issue), D611–613, 2005, http://www.mitomap.org.
7. Mueller, U.G. and Wolfenbarger, L.L., AFLP genotyping and fingerprinting, *Trends Ecol. Evol.* 14, 389–394, 1999.
8. Bagley, M.J., Anderson, S.L., and May, B., Choice of methodology for assessing genetic impacts of environmental stressors: polymorphism and reproducibility of RAPD and AFLP fingerprints, *Ecotoxicol.*, 10, 239–244, 2001.
9. D'surney, S.J., Shugart, L.R., and Theodorakis, C.W., Genetic markers and genotyping methodologies: an overview, *Ecotoxicol.* 10, 201–204, 2001.
10. Walsh, P.S., Fildes, N.J., and Reynolds, R., Sequence analysis and characterization of stutter products at the tetranucleotide repeat locus vWA, *Nucl. Acids Res.*, 24, 2807–2812, 1996.
11. Schumm, J.W., New approaches to DNA fingerprint analysis, *Promega Notes Mag.*, 58, 12–17, 1996.
12. Short Tandem Repeat DNA Internet Database, http://www.cstl.nist.gov/div831/strbase.

Forensic Canine STR Analysis

4

BURKHARD BERGER, PH.D., CORDULA EICHMANN, PH.D., AND WALTHER PARSON, PH.D.

Contents

4.1. Introduction

The dog has been part of human history longer than any other animal species and is deemed to be our closest companion and most popular pet. Therefore, from a forensic point of view, the dog can also be considered the most interesting animal species. As a consequence of the high abundance and close integration of dogs into human social life, forensically relevant cases involving dogs (such as accidents or dog attacks) are observed regularly. Forensic approaches to the identification of canines in fatal dog attacks have been published recently, demonstrating the relevance of this issue in forensic casework.[1–4] Additionally, dog hairs or dog saliva found at a crime scene can help by linking a crime scene and a potentially involved person, for example, the owner of the dog. This renders the analysis of canine DNA an important tool for forensic casework, complementing human DNA fingerprinting techniques.[2–6]

4.2 History and Domestication of the Dog

Morphological, behavioral, and genetic studies have shown that modern dogs (*Canis familiaris*) originated from the domestication of wild gray wolves (*Canis lupus*).[7–11] Results obtained mainly by the analysis of mitochondrial DNA (mtDNA) suggest there were several independent domestication events as indicated by different clades of mtDNA haplotypes and the genetic exchange between wolves and dogs that continued during coexistence over a wide geographical range.[7,10,12–15] The earliest archaeological remains of domestic dogs date back to about 10,000 to 15,000 years before the present. However, phylogenetic analyses of mtDNA data estimate an older time frame for the domestication, suggesting that dogs arose more than 100,000 years ago.[10]

The morphological and behavioral differences between dogs and wolves is the result of a strong human-mediated selection for desired morphological characters, behavioral traits, and/or the ability to learn and perform different tasks. Therefore, the dog population is divided into partially inbred genetic isolates called *breeds*. A number of different dog breeds have been bred with the intended purpose of guarding, hunting, herding, driving, pulling, etc. Today, more than 400 dog breeds are known to share people's homes. Most modern dog breeds are relatively young, with the majority having been developed within the last 300 years. Many of these were derived from a small number of original animals. Therefore, purebred dogs represent a limited genetic pool, with genetic characteristics derived from one or a small number of recent genetic founders. Compared to other animals, the population of dogs are characterized by a higher degree of isolation, narrower bottlenecks, and much better genealogical records. As a result of the extensive breeding, dogs are unique among mammalian species in the extent of variation they show in morphological traits such as height, weight, mass, shape, and behavior. No other mammalian species presents natural variation on a scale to rival dogs, yet individuals from nearly any breed can be mated to yield fertile offspring. This is an indication that the genetics of the domestic dog feature some remarkable aspects, which also have an impact on the principal strategy of forensic DNA typing of dogs.

4.3 The Canine Genome

The diploid chromosome number of the dog amounts to 76 autosomes plus two sex chromosomes. While the X chromosome is large, the Y chromosome is the smallest element in the karyotype. The dog genome has been extensively studied over the last 10 years and well-developed marker and genome maps are now available.[16–21] The most recent achievement concerns the sequencing of the dog genome,[22–24] and in July 2004, the first draft of the dog genome sequence has been deposited into free public databases by the National Human Genome Research Institute (http://www.broad.mit.edu/media/2004/doggenome_0714.html). This sequence is based on 7.6 times the coverage of the dog genome. The boxer breed was selected for this initial sequencing effort based on the lower variation rate in its genome relative to other breeds. The dog genome contains approximately 2.5 billion base pairs (Gb) of DNA. The estimate of the euchromatic genome size of approximately 2.5 Gb is similar to the estimated length of the mouse genome (2.5 Gb) but smaller than that of humans (2.9 Gb). Approximately 31% of the dog genomic sequence was identified as repetitive. This value is smaller than the content of known repetitive elements in human (46%) and mouse (38%) DNA. The availability of a high-quality genome sequence provides an important basis for an advanced understanding of the biology of dogs and their diseases. In addition, comparison of canine and human genomic sequences should allow a better understanding of the genetic basis of diseases affecting both humans and dogs. The identification of new informative polymorphisms (e.g., SNPs, microsatellites) will enhance our capability to study the genetic relationship of different breeds and to select genetic markers suitable for the identification of dog individuals, which can be used for forensic DNA typing.

4.4 Canine STRs in Forensic DNA Analysis

Today, the molecular information from highly variable short tandem repeat (STR) markers is being widely used to identify individual humans for forensic purposes (for an overview, see Jobling and Gill, 2004[25]). During the last 20 years, a number of human STR markers turned out to be very suitable for individualization, and as a consequence, these markers were investigated thoroughly (e.g., by sequencing of different alleles or by extensive population studies including individuals of different geographic origin). Additionally, an internationally accepted nomenclature for the designation of the alleles was established and allelic ladders as well as commercial kits have been developed, greatly unifying and facilitating the general use of this method. The techniques of STR analysis can also be used to identify animal individuals. As canine saliva or dog hairs can remain in any location where contact between dogs and humans has taken place, canine-specific STR analysis discloses a new approach for investigating dog attacks and other forensically important incidents involving dogs. Forensic approaches to investigate severe or fatal dog attacks are described[1,3,4] and individual dog bite injuries were reported in case studies, in which DNA profiling of hair, blood, or saliva of either the biting dog or the victim or both was successful. DNA profiling has complemented other methods for the investigation of fatal dog attacks, such as the examination of bite marks and the dog's mouth and stomach as well as pathological methods.[1,3,4,26,27]

In forensics — but also in many other research fields, such as behavioral ecology, conservation genetics, or paternity testing for breeding efforts — the genetic identification of animal individuals can be of prime importance. Because of the highly developed human DNA typing methods, the establishment of nonhuman STR analysis should be based on the human model.[28] The lessons that we have learned with human STR markers demonstrate that the forensic implementation of canine STR profiling requires the establishment of a nomenclature system for canine-specific STR markers that relates to the internationally accepted and generally adopted human STR allele nomenclature.[29–32]

4.5 Criteria for Forensic Canine STR Analysis

STR variation is widely used for investigating genetic traits of diseases, paternity testing, and for the analysis of the phylogenetic relationship of species or breeds. However, DNA analyses for forensic purposes have to take into account special aspects, such as minimal amounts of DNA, low DNA quality, mixtures, uniform terminology, and mutation probability. The choice of the genetic markers is extremely important, as it has consequences for all subsequent analyses. Ideal genetic markers should provide good DNA amplification, should be applicable for multiplex assays, should be easy to score (e.g., after capillary electrophoresis), and should exhibit high variability and heterozygosity. By choosing genetic markers with high heterozygosity, the total number of loci can be reduced in order to reach the desired probability of identity, as can the cost of the analysis and the risk of sample mix-up and contamination due to processing of multiple reactions.

Some loci provide enough PCR products when using large amounts of template DNA, but are difficult to optimize under extreme conditions with low amounts of template DNA or with degraded DNA. As such conditions

Table 1 Key Data of 15 Canine-Specific STRs. The Loci were Investigated[33–35] and are Listed According to Their HET Values. The Alleles have been Characterized by Sequence Analysis and a Repeat-Based Nomenclature was Introduced Earlier.[30] The Classification of the STR Loci (STR Class) Depends on the Complexity of the Polymorphic Region.[33,36] The Chromosomal Locations were Taken from the Literature. Heterozygosity (HETexp) and (HETobs), Polymorphism Information Content (PIC), and Probability of Identity (PI_{sib}) Values were Estimated as Described[34]

Marker	Repeat structure	STR Class	Chrom. location[37]	HET exp	HET obs.	PIC	PI_{sib}
ZUBECA6	$(GAAA)_n$	Complex	CFA5	0.949	0.888	0.947	0.277
FH2132	$(GAAA)_n$	Complex	CFA02	0.941	0.852	0.938	0.281
FH2087Ua	$(GAAA)_n$	Complex	CFA02	0.920	0.869	0.915	0.293
ZUBECA4	$(GAAA)_n$	Complex	CFA03	0.915	0.831	0.909	0.296
WILMS-TF	$(GAAA)_n$	Compound	CFA18[38]	0.897	0.832	0.889	0.306
PEZ15	$(GAAA)_n$	Compound	CFA16	0.887	0.759	0.877	0.312
PEZ6	$(GAAA)_n$	Complex	CFA27	0.881	0.842	0.870	0.316
FH2611	$(GAAA)_n(GGAA)_n(GAGA)_n$	Compound	CFA36	0.871	0.834	0.860	0.321
FH2087Ub	$(GAAA)_n$	Compound	CFA25	0.857	0.827	0.842	0.330
FH2054	$(GATA)_n$	Compound	CFA12	0.837	0.765	0.818	0.343
PEZ12	$(GAAA)_n$	Compound	CFA03	0.833	0.770	0.816	0.345
PEZ2	$(GGAA)_n$	Simple	no entry	0.782	0.644	0.749	0.379
FH2010	$(ATGA)_n$	Simple	CFA24	0.722	0.575	0.676	0.420
FH2079	$(GGAT)_n$	Simple	CFA24	0.713	0.322	0.665	0.426
VWF.X	$(AGGAAT)_n$	Simple	CFA27	0.614	0.539	0.558	0.494

are regularly found in forensic casework samples, the number of promising forensic STR markers is restricted. The purpose of this chapter is to elaborate on some important criteria for forensic DNA typing using canine STRs by using examples taken from our own research on canine STRs.[33–35,39] We investigated 15 highly polymorphic canine STR markers and two sex-related markers of 131 selected dogs, which were coamplified in PCR multiplex reactions (Table 1).

4.6 Length of the Repeat Unit

Dinucleotide microsatellites are very useful for many genetic analyses. For example, Parker et al.[40] successfully used a combination of SNPs and about 100 dinucleotide microsatellite markers to study the genetic relationships in a collection of 85 domestic dog breeds. However, tetra-, penta-, or hexanucleotide microsatellite loci are easier to score and show a reduced number

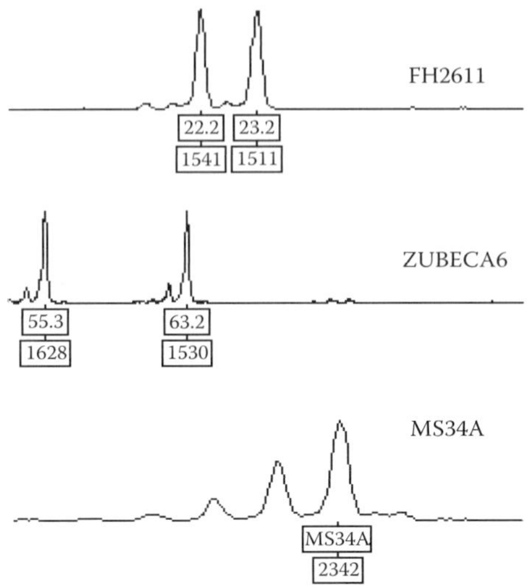

Figure 1 Genotyper plots of three canine-specific STR markers showing the different numbers and heights of stutter bands of tetrameric and dimeric STR markers. For FH2611 (tetra-repeat) the stutter peaks were lower than 10% of the allele peak. ZUBECA6 (tetra-repeat) showed the highest mean stutter peak height of all examined canine STRs (17%). For the Y-chromosomal marker MS34A (di-repeat) the mean stutter peak–allele peak ratios amounted to 57% and 22% for the n-2 and n-4 stutter, respectively.

and intensity of stutter peaks. Consequently, the use of tetranucleotide STRs considerably improves the consistency and reproducibility of microsatellite genotyping (Figure 1). Therefore, tetranucleotide STRs are the markers of choice in forensic applications.

The majority of canine STR markers described in the literature are based on di- and tetrameric repeats.[41–53] We selected as a set of potentially useful forensic markers tetrameric loci only (with the exception of the hexameric locus VWF.X) (Table 1). The investigated canine-specific STRs gave similar stutter band heights as those known from human-specific STRs. STR markers ZUBECA6 (17%) and FH2079 (13%) showed the highest mean stutter peak–allele peak ratios. For all other loci, the stutter peaks were lower than 10% of the main allele peak.

4.7 Repeat-Based Nomenclature

The majority of canine STR markers described in the literature are not yet characterized with respect to their sequence structure. Sequence analysis of the alleles is a requirement for the establishment of an STR nomenclature based on the number of repeat units.[29–32] Sequencing allows one to attain quantitative data with respect to the true nucleotide length of alleles and to detect sequence variants (e.g., alleles with identical fragment length but different nucleotide composition). In many published cases, the evidentiary material was investigated by means of canine-specific STR markers either with the aid of commercially available kits or by a selection of polymorphic canine STRs selected for population genetic studies.[2–5,37,54–56] These studies, as well as other population genetic investigations on dogs, have not used a uniform repeat-based nomenclature for the STR alleles.[2–6,37,54,56]

The lack of a uniform harmonized nomenclature makes the application of these markers difficult. Neither comparison between laboratories nor the establishment of frequency databases is possible. Mostly, the alleles were reported by the estimated fragment size as determined by electrophoresis of the PCR products. The designation of fragment lengths has a drawback in that data generated by one laboratory cannot be directly compared with the results of another laboratory, especially when different primers, alternative chemistry, and equipment (e.g., electrophoresis instruments) were used.

In order to provide the basis for the establishment of an STR nomenclature system based on the number of repeats, extensive sequencing of different alleles from our set of canine STR markers was performed (Figure 2). The nomenclature is adopted from the recommendations of the International Society of Forensic Genetics (ISFG) for the nomenclature of human STRs.[29–32]

Of the 15 investigated loci, four showed a simple repeat structure (Table 1). In these cases, no single-base-pair variants were observed.

Two other loci (FH2054, FH2087Ub) consisted of compound repeat units with only a few single-base-pair length variants and were still easy to interpret. The other nine loci showed a more complex polymorphic region partly including different repeat blocks and incomplete repeat units, which resulted in a relatively high portion of intermediate alleles. Of these, four loci (FH2087Ua, ZUBECA4, ZUBECA6, FH2132) showed an extremely complex repeat structure including di- to octomeric motifs as well as insertions of polyA-stretches within the repeat region (Figure 2).

The nomenclature for markers with a complex hypervariable polymorphic region deserves special attention and comment. An appropriate starting and end point of the repeat region were determined and the number of nucleotides therein were divided by four (assuming general tetrameric repeat structure) according to the recommendations proposed by Bär et al.[26] In some cases, however, this strategy can lead to predominantly intermediate alleles. For PEZ12, PEZ15, PEZ6, FH2087Ua, ZUBECA6, and FH2132, an arbitrarily shifted end point of the polymorphic region by one or two nucleotides minimizes the number of intermediate alleles and therefore simplifies allele calling.

Marker	Allele	(bp)	Repeat structure
FH2010	9	(224)	45bp **(ATGA)$_{9}$** 143bp
FH2079	5	(266)	154bp **(GGAT)$_{5}$** 92bp
PEZ2	8	(109)	32bp **(GGAA)$_{8}$** 45bp
VWF.X	9	(157)	70bp **(AGGAAT)$_{9}$** 33bp
FH2054	11	(151)	46bp **(GATA)$_{11}$** 61bp
	12	(155)	46bp **(GATA)$_{8}$ GTTA(GATA)$_{3}$** 61bp
FH2087Ub	7	(227)	58bp **(GAAA)$_{7}$** 141bp
	7	(227)	58bp **(GAAA)$_{6}$ GCAA** 141bp
FH2611	14	(178)	69bp **(GAAA)$_{7}$ (GGAA)$_{3}$(GAGA)$_{4}$** 53bp
	19.2	(200)	69bp **(GAAA)$_{11}$(GGAA)$_{5}$(GAGA)$_{3}$(GA)** 53bp
WILMS-TF	12	(283)	68bp **(GAAA)$_{12}$** 167bp
	15.1	(296)	68bp **(GAAA)$_{3}$G(GAAA)$_{12}$** 167bp
	13.3	(290)	68bp **(GAAA)$_{10}$GAA(GAAA)$_{3}$** 167bp
PEZ12	12	(267)	84bp **(GAAA)$_{10}$GAAGAAAG** 135bp
	21.2	(305)	84bp **(GAAA)$_{2}$GAA(GAAA)$_{11}$GAA(GAAA)$_{3}$GAA(GAAA)$_{3}$G** 135bp
PEZ15	13	(211)	109bp **(GAAA)$_{12}$GAAG** 50bp
	18.2	(233)	109bp **(GAAA)$_{4}$GAA(GAAA)$_{2}$GAA(GAAA)$_{9}$GAAGAAAG** 50bp
	13.3	(214)	109bp **(GAAA)$_{4}$GAA(GAAA)$_{8}$GAAG** 50bp
PEZ6	26	(185)	46bp **AAAAAAAGAAGAAAGAAGGAAAGA(GAAA)$_{2}$AAGAAAGAGAAAGAA(GAAA)$_{14}$ A** 35bp
	23.3	(176)	46bp **AAAAAAAAGAA(GAAA)$_{2}$GAAGAAAGAAGGAAAGA(GAAA)$_{2}$AA(GAAA)$_{12}$ A** 35bp
FH2087Ua	32	(357)	98bp**(GAAA)$_{15}$GAA(GAAA)$_{2}$AAAA(GAAA)$_{11}$GAA(GA)$_{2}$CA** 131bp
	37.3	(380)	98bp**(GAAA)$_{13}$AA(GAAA)$_{4}$GAA(GAAA)$_{3}$AAA(GAAA)$_{2}$GA(GAAA)$_{11}$G(GA)$_{3}$CA** 131bp
ZUBECA4	38	(415)	105bp **(GAAA)$_{6}$GAA(GAAA)$_{2}$GAAGAAAAA(GAAA)$_{3}$AAAAA(GAAA)$_{2}$(GAAAA)$_{3}$AA(GAAA)$_{12}$GA(GAAA)$_{4}$** 158bp
	46.2	(449)	105bp **GAA(GAAA)$_{8}$GAA(GAAA)$_{2}$GAAGAAAAA(GAAA)$_{3}$AAAAA(GAAA)$_{3}$AGAAAAAA(GAAA)$_{4}$GA(GAAA)$_{19}$** 158bp
ZUBECA6	45	(308)	73bp **(GA)$_{12}$(GAAA)$_{4}$A(GAAA)$_{2}$G(GAAA)$_{2}$GAA(GAAA)$_{2}$GA(GAAA)$_{12}$(GGAA)$_{14}$GAAAG** 55bp
	58.2	(362)	73bp **(GA)$_{3}$GAAA(GA)$_{4}$(GAAA)$_{2}$G(GAAA)$_{2}$GAAGA(GAAA)$_{2}$GAA(GAAA)$_{16}$(GGAA)$_{12}$GAAAGA(GAAA)$_{16}$G** 55bp
	53.3	(343)	73bp **(GA)$_{3}$GAAA(GA)$_{7}$(GAAA)$_{3}$A(GAAA)$_{2}$G(GAAA)$_{2}$GAAGAAAGA(GAAA)$_{2}$GAA(GAAA)$_{6}$(GGAA)$_{16}$(GAAA)$_{13}$G** 55bp
	58.3	(363)	73bp **(GA)$_{3}$GAAA(GA)$_{4}$(GAAA)$_{3}$A(GAAA)$_{2}$G(GAAA)$_{2}$GAAGA(GAAA)$_{5}$(GGAA)$_{19}$GA(GAAA)$_{18}$GAA(GAAA)$_{2}$G** 55bp
FH2132	36	(251)	69bp **AAAAAGAAAGAA(GAAA)$_{13}$(GGAA)$_{11}$(GAAA)$_{2}$AGAA(GAAA)$_{5}$AAAG** 38bp
	44.2	(285)	69bp **AAAAAGAAAGAA(GAAA)$_{3}$GA(GAAA)$_{11}$(GGAA)$_{18}$(GAAA)$_{2}$AGAA(GAAA)$_{5}$AAAG** 38bp
	58	(339)	69bp **AAAAAGAAAGAAGAAAA(GAAA)$_{10}$(GGAA)$_{15}$(GAAA)$_{4}$A(GAAA)$_{9}$A(GAAA)$_{14}$AAAAG** 38bp
	51.3	(314)	69bp **AAAAAGAAAGAAGAAAA(GAAA)$_{13}$(GGAA)$_{7}$(GAAA)$_{3}$AGAA(GAAA)$_{11}$A(GAAA)$_{11}$AAAAG** 38bp
	40.1	(268)	69bp **AAAAAAAA(GAAA)$_{12}$(GGAA)$_{10}$(GAAA)$_{2}$AGAA(GAAA)$_{12}$AAAAG** 38bp
	41.2	(273)	69bp **AAAAAAAA(GAAA)$_{7}$A(GAAA)$_{5}$(GGAA)$_{12}$(GAAA)$_{2}$AGAA(GAAA)$_{11}$AAAAG** 38bp

Figure 2 Sequence data of five selected canine-specific STR markers. Depending on the complexity of the marker, representative numbers of sequence variants are shown for each locus.

4.8 Allele Frequencies, Heterozygosity, and PIC Values

The relative frequencies of five selected canine STR loci in our population sample (n = 131) are shown in Figure 3. Heterozygosity is a widely used measure for the genetic information of a marker. Generally, a good forensic marker should exhibit high heterozygosity, ranging between 0.6 and 0.8. The HET(exp) and HET(obs) values as well as the polymorphism information content (PIC) values of some selected canine STRs are depicted in Table 1. Usually, the expected and observed HET values were similar with the exception

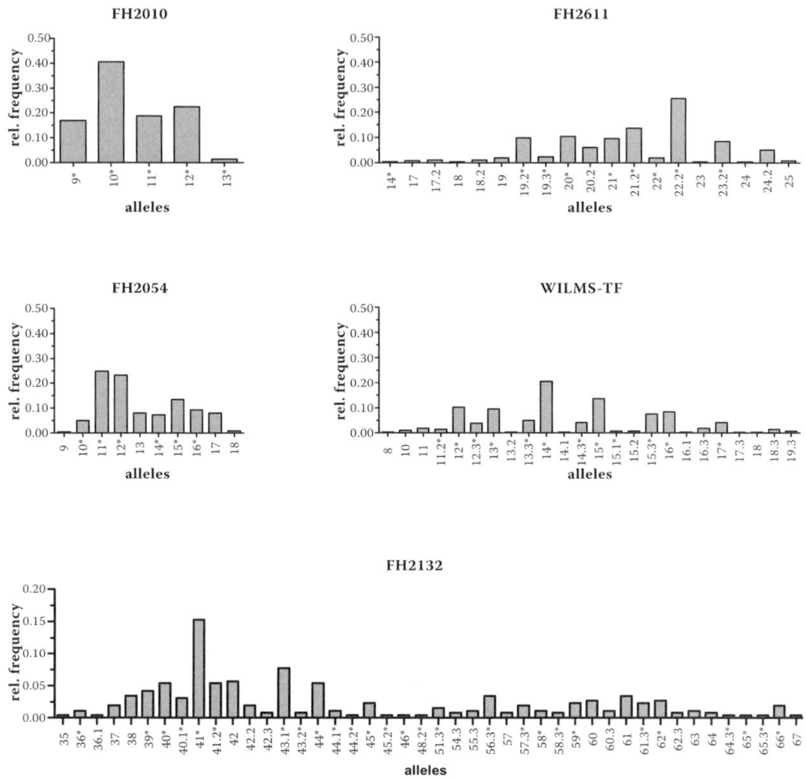

Figure 3 Allele frequency distributions of five selected canine-specific STR markers: FH2010 (simple repeat structure), FH2054, FH2611, and WILMS-TF (complex STRs) and FH2132 (hyper-variable STR). The classification of the STR loci depends on the complexity of the polymorphic region according to Urquhart et al.[36] and is described in more detail[33] (*): sequenced alleles.

of FH2079 (HET[exp] = 0.7, HET[obs] = 0.3), where we observed elevated homozygosity. The marker ZUBECA6 brought the highest HET values (Table 1). The markers with the lowest HET values feature low numbers of alleles ($n < 10$) and unequal distributions of allele frequencies (for example, see marker FH2010 in Figure 3). Further, they show one common allele each with a frequency of at least 30%.

4.9 Multiplexing of Canine STRs — Examples

In order to reduce the costs and the time to carry out an analysis on a series of STR markers, a single-tube PCR reaction that amplifies multiple STR loci ("multiplexes") was developed, and today the majority of human forensic casework is now done using commercially developed autosomal STR multiplexes.[25] Therefore, the development of highly discriminating canine STR multiplexes is an important step toward the implementation of canine DNA fingerprinting.

Theoretically, the best approach is to combine only the most polymorphic markers, which on the other hand are more likely to cause technical problems for allele calling due to the increased occurrence of sequence and allele variants.[33] Simple repeat loci and some compound loci may be the markers of choice for routine forensic casework, as they are easy to type and display short fragment lengths, a fact that is important for the analysis of degraded DNA. The complex markers might be useful in individual cases due to their excellent variability. However, they are difficult to characterize and usually display long fragment lengths, which might limit their use to high quality DNA only.

As shown in Table 2, our canine-specific set of STR loci and two sex-related markers[57] were coamplified in three PCR multiplexes (MP1-3). An in-depth description of the three PCR multiplexes is given.[34] The amplification of ZUBECA6 in MP1 and MP2 as well as of FH2611 in MP2 and MP3 served as an internal control. The three MPs were predominantly used with a DNA amount of 2 nanograms. The multiplexes were also tested for their sensitivity with DNA templates consisting of 100 pg, 250 pg, 500 pg, and 1 ng DNA. MP1 and MP3 brought full profiles with 100 pg of template DNA, for MP2 the lower detection limit was 250 pg. For a new multiplex (designated MP4), PCR primers for the markers FH2087Us, PEZ 12, and WILMS-TF were redesigned in order to reduce the length of the amplification product compared with the design published earlier.[43,57,58] Careful PCR primer design that introduces short amplicons increases the chance of a successful STR analysis of casework samples. The new amplification primers can be applied to degraded casework samples or samples having just a small amount of DNA. Figure 4 shows an electropherogram of 2 nanograms canine DNA amplified with the MP4 short multiplex.

Table 2 Canine-Specific STR Markers Used for the PCR Multiplexes Described in the Text. MP-4 Short Includes Redesigned PCR Primers for Three Markers (*) to Reduce the Length of the Amplification Products

PCR Multiplex	Canine STR Marker
MP1	VWFX, FH2087U, PEZ2, PEZ15, ZUBECA6, SRY and CHR.X
MP2	FH2611, FH2079, PEZ6, FH2132, ZUBECA6 and ZUBECA4
MP3	PEZ12, FH2611, FH2054, FH2010 and WILMSTF
MP4short	FH2087Us*, PEZ15, PEZ12*, WILMS-TF*

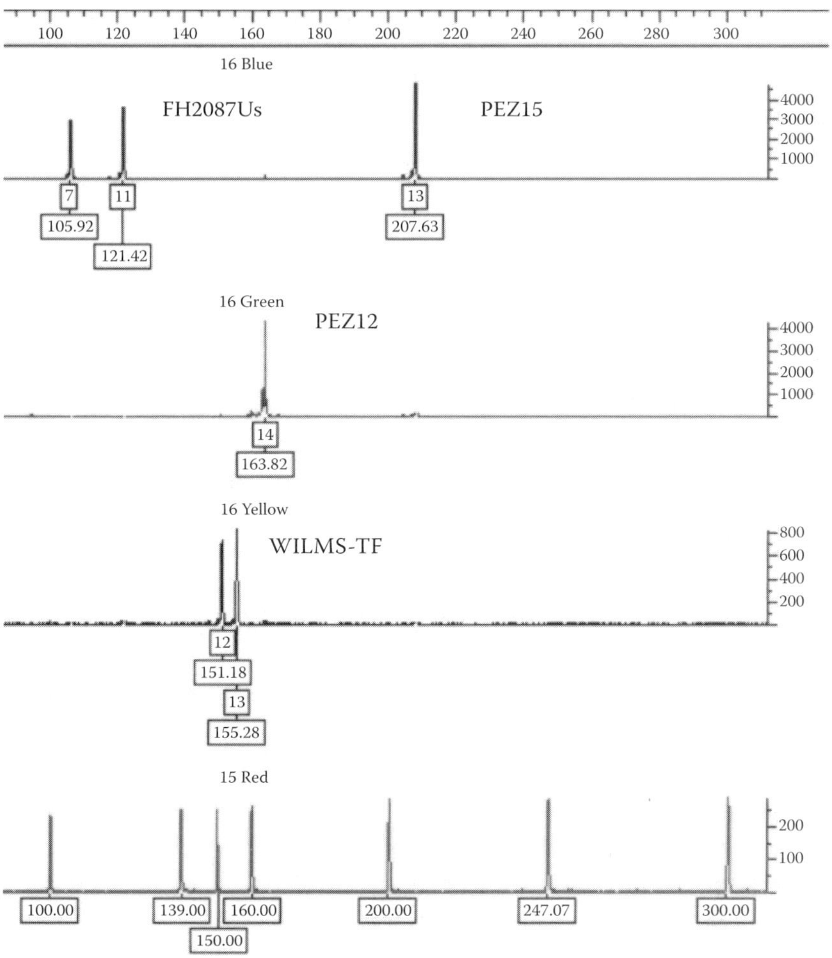

Figure 4 Genotyper plot of the amplification of a dog sample using a canine-specific PCR multiplex consisting of four STRs (MP-4 short). A new primer design was applied for these markers in order to shorten the amplification length. Peaks are labeled with respect to a repeat-based nomenclature.[33]

4.10 Canine Specificity

For forensic canine DNA analyses, the impact of contaminant human DNA has to be considered. Therefore, mixtures of known human and canine DNA and of various amounts were prepared in order to (1) verify the canine specificity of the canine STR markers used, (2) test for the possible formation of artifacts when human DNA is amplified with the canine specific PCR primers, and (3) attempt to do both canine and human STR analysis for identification of both involved individuals, simultaneously.

The amplifications of up to 100 ng of human DNA with the three canine multiplexes resulted in no artifactual peaks confirming the specificity of the primers used. Likewise, amplifying of up to 2 nanograms of canine DNA with a human-specific PCR multiplex (AmpF*l*STR SGM Plus PCR Amplification Kit, Applied Biosystems) produced no artifacts. However, 100 ng of canine DNA amplified with the human-specific multiplex showed a peak corresponding to approximately 99 base pairs. This phenomenon has already been described[27] and does not negatively influence the allele calling. The results of the mixture study suggest clearly that an unambiguous STR profiling of canine DNA samples is possible even if a high background of human DNA is present. Additionally, human STR analysis is possible in the presence of canine DNA.

4.11 Individual Identity and Population Structure

In forensic casework, multilocus genotyping of canine STR markers is done with the aim of identifying dog individuals. As a consequence, it is necessary to quantify the power of the different markers (of the marker set) to resolve different individuals. Furthermore, this should allow an estimate of how many canine STR markers would be suitable for forensic applications.

For human DNA testing, the weight of the DNA evidence is most commonly reported in terms of the match probability (Pm) or probability of identity (PI) — the chance of two unrelated people sharing a profile. For independently inherited loci, Pm is calculated by multiplying the individual allele frequencies in the profile in question ("product rule"). However, there are a number of situations in which Pm can be substantially increased and the use of the product rule without adjusted allele proportions has not been applied without controversy and adaptations.[25,59,60] Particularly for nonhuman populations, a series of circumstances can cause frequencies of alleles to vary between subpopulations. Nonindependence among alleles and loci can be found especially within small populations (including historic "population bottlenecks") or species with a complex social structure caused by shared ancestry within family clans. Thus, in many natural populations the theoretical equations for PI (assuming large populations and no shared ancestry) will underestimate the true probability of finding identical genotypes. Waits, Luikart, and Taberlet[61] demonstrated the potential inaccuracies of using standard PI formulas to determine the probability of obtaining matching multilocus genotypes when two samples are drawn at random from natural animal populations. The risk of incorrectly assigning a match based on the PI is potentially higher for populations of social species in which many related individuals will probably be sampled and in populations that violate random mating principles.[61]

The history of the domestic dog, and especially of the different dog breeds, suggests that nonrandom mating (selection due to breeding efforts) as well as population bottlenecks can be regarded as a feature of the modern dog population. SNP data demonstrate the reality of breed barriers between dog breeds and show a much higher level of genetic differentiation of the dog population compared to humans.[9,40] Additionally, the linkage disequilibrium in dog breeds was found to be 20–100 times more extensive than in humans.[62] Consequently, we have to take into account the bias due to the structure of the dog population by using formulas that include parameters that estimate population substructure and the fraction of siblings. Therefore, we are of the opinion that formulas developed for genetic studies of wild animals (e.g., concerning behavioral ecology or conservation genetics) could be suitable for PI estimates within dog populations. In the following, we discuss our genetic data drawn from a population sample of 131 dogs by applying the equation for the PI among siblings (PI_{sib}) suggested by Taberlet and colleagues.[61,62] The PI_{sib} is based on an equation for codominant loci[64] and provides an upper bound for identification estimations in natural animal populations. The PI_{sib} values were computed individually for each locus using allele frequencies obtained from population data[33] and are listed in Table 1. Additionally, the PI_{sib} was computed for different sets of loci. For the combination of all 15 loci, the PI_{sib} amounted to 8.5×10^{-8}. A goal of our study was to estimate how many canine STR markers would be necessary to attain a PI suitable for forensic applications. Generally, low PI values would be desirable, but for practical, technical, and economic reasons, feasible limits have to be discussed. A PI_{sib} between 0.001 and 0.0001 is deemed to be sufficiently low for the identification of individuals in natural animal populations.[56] This is in the range of probability values recommended for reporting human STR matches, including siblings.[55] In order to assess the number of loci required for attaining the proposed PI_{sib} range, we sequentially multiplied the single PI_{sib} values of the markers starting with (1) the lowest and (2) the highest HET value (Table 1). When markers with increasing HET values were combined, a selection of 10 markers was necessary to achieve a PI_{sib} value of approximately 0.0001, whereas the inverse order required only eight loci. An estimated minimum of eight to ten canine STR loci should be considered, which is in agreement with the conclusions drawn by Waits, Luikart, and Taberlet[61] and DeNise et al.[65]

4.12 Sampling Strategies for Population Data

A crucial point is whether the genotypes of a case sample are compared to results obtained from a proper control population, because a difference in allele and genotype frequencies between the (sub)population from which the case sample originates and the control population leads to distorted results for calculating random match probabilities. Therefore, proper sampling strategies are of decisive importance. Depending on the aim of a study, different sampling strategies can be found in the case of dog population evaluations. Some studies focus on genetic differences between breeds.[40,66] In these cases, a proper number of individuals per breed were collected and examined. In other studies, particularly rare breeds or breeds of a certain geographic origin were sampled.[67] However, these sampling strategies do not allow for drawing a picture of the whole genetic variation possible in a dog population, as the proportions of different breeds as well as the fraction of crossbreeds are not addressed. The composition of purebred dogs within a population can be rather complex. For example, the approximately 90,000 purebred dogs registered by the American Kennel Club are sorted into over 150 distinct breeds, whereas the most popular 20 breeds account for 70% of all individuals (see www.akc.org and Sutter and Ostrander[9]). Additionally, the fraction of crossbreeds amounts to nearly 50% of the total population. For forensic questions, with a goal of identifying dog individuals, the whole dog population has to be regarded, and we think that the most efficient way to get an image of the population structure is by random sampling. Diverse sampling designs are possible, but so far there is a lack of forensic studies dealing with this topic. As the better part of dog samples comprising the data presented here were received from the routine work of a veterinarian, we think that our preliminary dataset largely reflects the real population of dogs of our investigation area. However, we know that this is just the first step toward a comprehensive forensic sampling strategy for collecting population data of the domestic dog and further discussions dealing with this topic are needed. Independent from these considerations, increasing the sample size will lead to a sustainable improvement of allele frequency estimates and is therefore an additional paramount aim of future studies.

4.13 Canine STRs in Casework: Analysis of Dog Bites

The forensic relevance of dog bite cases arise mainly from severe and/or mortal injuries involving stray dogs or owned dogs that are allowed to run free. In these cases, an unambiguous identification of the biting dog is vitally important.

The social significance of severe dog attacks can be substantiated by some statistical facts: (1) in the United States more than 3.5 million bites per year are estimated, (2) in Germany every year 30,000 dog bite injuries are registered, and (3) in Hong Kong 200 dog bites are recorded every month.[68] In the United States, there were 238 deaths reported from 1979 to 1998 and in Germany, 55 deadly dog bite injuries were registered from 1968 to 2002.[1,69]

The dog bite death rates in Australia and the United States range between 0.004 and 0.07 per 100,000 inhabitants.[70] The injury rate was highest for children aged 5–9 years and decreased with increasing age; the rate was higher for boys than for girls. Because of their height, children are more likely to be bitten on the face, neck, and head than older persons, whereas most injuries to adult humans affect the extremities.[68,70–73] For persons older than 15 years, the differences between the rate for males and females was not statistically significant.[68,70–78] In most cases of canine aggression toward children, misinterpretations of meaning and message could be the cause for injuries. Just as humans can misinterpret a wagging tail, dogs can misinterpret a screaming child. Children may be uncoordinated and may appear unpredictable to dogs because of their sudden shifts in postures and vocal range when excited.[68] Although the bulk of dog bite victims were bitten by their own dogs, bites inflicted by stray dogs or owned dogs that are allowed to run free are common.[68] The German Shepherd, rottweiler, and pit bulls, as well as their crossbreeds, lead the statistics of dog breeds involved in bite injuries.[68,70,71,75,77]

Several dog bite injuries were reported in casework in which DNA profiling of hair, blood, or saliva of either the biting dog or the victim or of both involved individuals was successful, complementing other methods for investigating fatal dog attacks, such as examining the bite marks sustained by the victim, examination of the dog's stomach and mouth, and other pathological and forensic methods.[1,3,4,26,27]

The analysis of canine DNA from swabs and bandages directly from the wound and its surrounding area will be of stronger evidence concerning the identity of the perpetrator compared with other means of evidence such as hair samples or blood and saliva stains found on the clothes of the victim. Dog bites assume intensive contact between the dog's muzzle and the human skin. As the human skin is a soft tissue, it is unlikely that the dog hurts himself during the biting of a person. Even then, it cannot be completely excluded that in some cases canine blood remains on the bite. However, we presume that the canine material left on the wound is in most cases canine saliva. The remains of canine saliva collected directly from the bite or the wound bandage need to be suitable for canine STR profiling. As observed in human DNA analysis, saliva has been recovered and successfully analysed from various substrates, including human skin and human bite marks.[80,81] Canine saliva is best recovered from the area directly surrounding the human wound, where the dog's gums and flews contacted the skin of the bitten person. Severe injuries, besides being more relevant to forensic investigations, will probably also contain more canine material. However, the wound bandages of such cases are very likely to contain mixtures with respective quantities of human blood, which potentially has adverse effects on the ability to reconstruct the canine-specific STR profile.

In order to investigate canine-specific STRs on dog bite marks in a systematic approach, we used swabs and wound bandages of 52 dog attack cases received from the routine casework of the Department of Trauma Surgery for extraction and analysis of the canine-specific STRs. Sample collection proved to be demanding because medical first aid was of paramount importance and data privacy made case-specific information not available. However, these limitations did not negatively influence the predominant methodical background of this study, namely to test canine STR profiling using samples collected directly from the dog bite itself.

The three PCR multiplexes MP1-3 (as described above) were used for DNA profiling (for details, see Eichmann et al.[34]). The samples consisted of bandages presumably used immediately after the attack to stop the blood flow, which were then removed by the medics on admission of the patient, and of swabs taken from the area surrounding the wound. They were grouped according to the amount of visible blood into the following three classes: (1) large amounts of blood (B++), (2) medium quantities of blood (B+), and (3) no visible blood (B–).

Overall, 30.8% of the samples gave a full canine-specific STR profile, 9.6% gave a partial profile, and 59.6% showed no results. Additionally, there seemed to be a correlation between the amount of blood — most likely of human origin — on the collected material and the ability to retrieve a canine-specific STR profile. Bandages and/or swabs with a large amount of visible blood stains (B++) produced full canine profiles in 65% of samples tested, samples belonging to the (B+)-class brought useful results in 24% of samples, and no canine STR profiles were obtained from samples belonging to the (B–) class. As shown in Figure 5, this trend becomes more apparent by grouping the data into the three categories — "full profile," "partial profile," and "no result." However, it cannot be excluded that the wounds were treated by the victims prior to the examination at the hospital. This may especially be true for the less severe attacks, as no severe wounding had been involved and the victims may have cleaned the surface, thus removing dog saliva prior to medical aid. This would explain why swabs and bandages without visible blood stains (B–) were usually less successful in terms of identifying canine-specific DNA compared with severe attacks involving relatively high amounts of human blood. Even these high concentrations of human DNA do not seem to interfere with successful canine STR typing. The positive correlation between the amount of human blood and canine saliva surrounding the wound — as a consequence of the intensive contact — supports the assumption that severe injuries provide a better chance of obtaining a successful canine-specific DNA result than more harmless dog bites.

Summarizing canine-specific DNA profiling of dog bite marks proved to be a promising way of investigating forensic accidents involving dogs. Our

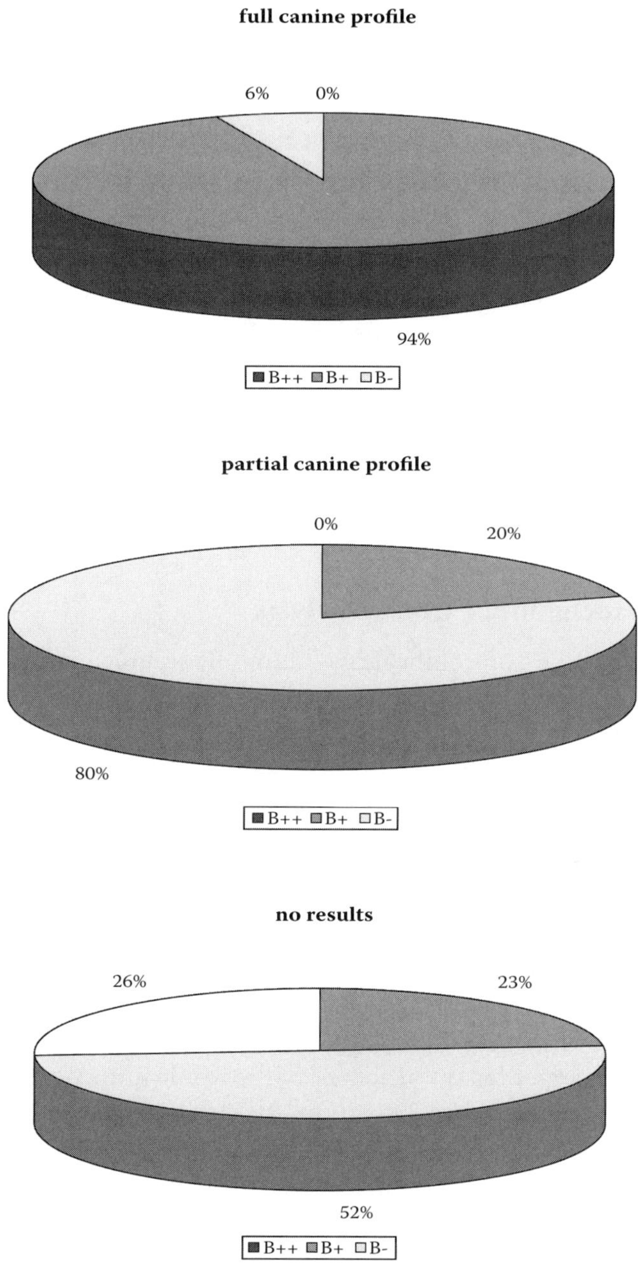

Figure 5 Success rate of canine-specific STR profiling depending on the amount of visible blood on the sampling device. (**See color insert after p. 108.**) The three categories "full profiles," "partial profiles," and "no results" are distinguished. The 52 samples were grouped into three classes: (B++), much blood (n = 23); (B+), little blood (n = 21); (B–), no blood (n = 8).

results suggest that in dog biting cases where the identity of the offending dog is in question, both wound bandages and swabs serve as useful stain material in order to perform canine-specific STR analysis.

4.14 Additional Methods for Canine DNA Analysis

4.14.1 X- and Y-Chromosomal Markers

For sex determination, loci located on the X- and Y-chromosome can be used. For example, we coamplified in MP1 the gender-specific markers CHR.X and SRY. Similar to the human-specific gender test, the profile of a female dog shows a single CHR.X peak, whereas the profile of a male dog additionally includes the SRY peak. Sex determination with these two markers was successful in all cases. In our population sample, 61 male (47%) and 70 female (53%) dogs were identified by a veterinarian. Sex determination by means of CHR.X and SRY confirmed these findings in all cases.

4.14.2 Mitochondrial DNA Analysis

Shed human hairs are one of the most commonly secured biological evidence materials at crime scenes. As single shed hairs and hair shafts contain only minute amounts of undegraded DNA, mitochondrial DNA, which is present in more than 1000 copies per cell, is a suitable target for analysis giving a higher probability of finding intact DNA copies in samples that contain small amounts of DNA.[82] However, a drawback is that the exclusion capacity is lower for mtDNA than for nuclear DNA since the mtDNA molecule is small and maternally inherited, and does not recombine. Two regions of the mtDNA are of main importance for forensic testing of nonhuman DNA: (1) the analysis of the cytochrome b (cytb) gene has been applied to the identification of vertebrate species[83] and (2) the analysis of the control region (CR) is performed for further differentiation.[67,82,84–87]

Kim et al.[88] presented the sequence of the whole mitochondrial genome, and recent efforts have been made to standardize nomenclature for the mtDNA control region of the dog.[89] The mtDNA sequence variation is more limited in domestic dogs than in humans.[84] The lower variation among dogs can be explained by the limited number of animals that was involved in the domestication of the wolf,[10] and the limited amount of time that has passed since that event. Furthermore, although mitochondrial DNA analyses have been used successfully to elucidate the relationship between the domestic dog and the wolf, the evolution of mitochondrial DNA is too slow to allow inference of relationship among modern dog breeds, most of which have existed as closed breeding populations for fewer than 400 years.[9]

References

1. De Munnynck, K. and Van de Voorde, W., Forensic approach of fatal dog attacks: a case report and literature review, *Int. J. Legal Med.*, 116, 295–300, 2002.
2. Muller, S., Flekna, G., Muller, M., and Brem, G., Use of canine microsatellite polymorphisms in forensic examinations, *J. Hered.*, 90, 55–56, 1999.
3. Padar, Z., Egyed, B., Kontadakis, K., Furedi, S., Woller, J., Zoldag, L. et al., Canine STR analyses in forensic practice. Observation of a possible mutation in a dog hair, *Int. J. Legal Med.*, 116, 286–288, 2002.
4. Shutler, G.G., Gagnon, P., Verret, G., Kalyn, H., Korkosh, S., Johnston, E. et al., Removal of a PCR inhibitor and resolution of DNA STR types in mixed human-canine stains from a five year old case, *J. Forens. Sci.*, 44, 623–626, 1999.
5. Ichikawa, Y., Takagi, K., Tsumagari, S., Ishihama, K., Morita, M., Kanemaki, M. et al., Canine parentage testing based on microsatellite polymorphisms, *J. Vet. Med. Sci.*, 63, 1209–1213, 2001.
6. Padar, Z., Angyal, M., Egyed, B., Furedi, S., Woller, J., Zoldag, L. et al., Canine microsatellite polymorphisms as the resolution of an illegal animal death case in a Hungarian zoological gardens, *Int. J. Legal Med.*, 115, 79–81, 2001.
7. Leonard, J.A., Wayne, R.K., Wheeler, J., Valadez, R., Guillen, S., and Vila, C., Ancient DNA evidence for Old World origin of New World dogs, *Science*, 298, 1613–1616, 2002.
8. Olsen, S.J., *Origins of the Domestic Dog: The Fossil Record,* University of Arizona Press, Tucson, 1985.
9. Sutter, N.B. and Ostrander, E.A., Dog star rising: the canine genetic system, *Nat. Rev. Genet.*,5, 900–910, 2004.
10. Vila, C., Savolainen, P., Maldonado, J.E., Amorim, I.R., Rice, J.E., Honeycutt, R.L. et al., Multiple and ancient origins of the domestic dog, *Science,* 276, 1687–1689, 1997.
11. Wayne, R.K., Molecular evolution of the dog family, *Trends Genet.*, 9, 218–224, 1993.
12. Flagstad, O., Walker, C.W., Vila, C., Sundqvist, A.K., Fernholm, B., Hufthammer, A.K. et al., Two centuries of the Scandinavian wolf population: patterns of genetic variability and migration during an era of dramatic decline, *Mol. Ecol.* 12, 869–880, 2003.
13. Vila, C., Maldonado, J.E., and Wayne, R.K., Phylogenetic relationships, evolution, and genetic diversity of the domestic dog, *J. Hered.*, 90, 71–77, 1999.
14. Vila, C., Walker, C., Sundqvist, A.K., Flagstad, O., Andersone, Z., Casulli, A. et al., Combined use of maternal, paternal and bi-parental genetic markers for the identification of wolf-dog hybrids, *Heredity*, 90, 17–24, 2003.
15. Wayne, R.K. and Ostrander, E.A., Origin, genetic diversity, and genome structure of the domestic dog, *Bioessays*, 21, 247–257, 1999.

16. Breen, M., Langford, C.F., Carter, N.P., Holmes, N.G., Dickens, H.F., Thomas, R. et al., FISH mapping and identification of canine chromosomes, *J. Hered.*, 90, 27–30, 1999.

17. Breen, M., Bullerdiek, J., and Langford, C.F., The DAPI banded karyotype of the domestic dog (*Canis familiaris*) generated using chromosome-specific paint probes, *Chromos. Res.*, 7, 401–406, 1999.

18. Breen, M., Jouquand, S., Renier, C., Mellersh, C.S., Hitte, C., Holmes, N.G. et al., Chromosome-specific single-locus FISH probes allow anchorage of an 1800-marker integrated radiation-hybrid/linkage map of the domestic dog genome to all chromosomes, *Genome Res.*, 11, 1784–1795, 2001.

19. Breen, M., Hitte, C., Lorentzen, T.D., Thomas, R., Cadieu, E., Sabacan, L. et al., An integrated 4249 marker FISH/RH map of the canine genome, *BMC Genom.*, 5, 65, 2004.

20. Guyon, R., Lorentzen, T.D., Hitte, C., Kim, L., Cadieu, E., Parker, H.G. et al., A 1-Mb resolution radiation hybrid map of the canine genome, *Proc. Natl. Acad. Sci. USA*, 100,5296–5301, 2003.

21. Klukowska, J., Szczerbal, I., Rickli, O., Switonski, M., Dolf, G., and Schelling, C., Seven bacterial artificial chromosome-derived canine microsatellite-linking physical and genetic maps, *Anim. Genet.*, 35, 252–253, 2004.

22. Kirkness, E.F., Bafna, V., Halpern, A.L., Levy, S., Remington, K., Rusch, D.B. et al., The dog genome: survey sequencing and comparative analysis, *Science*, 301, 1898–1903, 2003.

23. Lindblad-Toh, K., Wade, C.M., Mikkelsen, T.S., Karlsson, E.K., Jaffe, D.B., Kamal, M. et al., Genome sequence, comparative analysis and haplotype structure of the domestic dog, *Nature*, 438(8), 803–819, 2005.

24. Ostrander, E.A., Wayne, R.K., The canine genome, *Genome Res.*, 15, 1706–1716, 2005.

25. Jobling, M.A. and Gill, P., Encoded evidence: DNA in forensic analysis, *Nat. Rev. Genet.*, 5, 739–751, 2004.

26. Brauner, P., Reshef, A., and Gorski, A., DNA profiling of trace evidence—mitigating evidence in a dog biting case, *J. Forens. Sci.*, 46, 1232–1234, 2001.

27. Wiegand, P., Schmidt, V., and Kleiber, M., German shepherd dog is suspected of sexually abusing a child, *Int. J. Legal Med.*, 112, 324–325, 1999.

28. Budowle, B., Garofano, P., Hellmann, A., Ketchum, M., Kanthaswamy, S., Parson, W. et al., Recommendations for animal DNA forensic and identity testing, *Int. J. Legal Med.*, 119(5), 295–302, 2005.

29. Bär, W., Brinkmann, B., Budowle, B., Carracedo, A., Gill, P., Lincoln, P. et al., DNA recommendations. Further report of the DNA Commission of the ISFH regarding the use of short tandem repeat systems, International Society for Forensic Haemogenetics, *Int. J .Legal Med.*, 110, 175–176, 1997.

30. Gill, P., Kimpton, C., d'Aloja, E., Andersen, J.F., Bar, W., Brinkmann, B. et al., Report of the European DNA profiling group (EDNAP)—towards standardisation of short tandem repeat (STR) loci, *Forens. Sci. Int.* 65, 51–59, 1994.

31. Gill, P., Brinkmann, B., d'Aloja, E., Andersen, J., Bar, W., Carracedo, A. et al., Considerations from the European DNA profiling group (EDNAP) concerning STR nomenclature, *Forens. Sci. Int.* 87, 185–192, 1997.

32. Gill, P., Brenner, C., Brinkmann, B., Budowle, B., Carracedo. A., Jobling, M.A. et al., DNA Commission of the International Society of Forensic Genetics: recommendations on forensic analysis using Y-chromosome STRs, *Int. J. Legal Med.*, 114, 305–309, 2001.

33. Eichmann, C., Berger, B., and Parson, W., A proposed nomenclature for 15 canine specific polymorphic STR loci for forensic purposes, *Int. J. Legal Med.*, 118, 249–266, 2004.

34. Eichmann, C., Berger, B., Steinlechner, M., and Parson, W., Estimating the probability of identity in a random dog population using fifteen highly polymorphic canine STR markers, *Forens. Sci. Int.*, 151, 37–44, 2005.

35. Eichmann, C., Berger, B., Reinhold, M., Lutz, M., and Parson, W., Canine-specific STR typing of saliva traces on dog bite wounds, *Int. J. Legal Med.* 118, 337–342, 2004.

36. Urquhart, A., Kimpton, C.P., Downes, T.J., and Gill, P., Variation in short tandem repeat sequences—a survey of twelve microsatellite loci for use as forensic identification markers, *Int. J. Legal Med.*, 107, 13–20, 1994.

37. Padar, Z., Egyed, B., Kontadakis, K., Zoldag, L., and Fekete, S., Resolution of parentage in dogs by examination of microsatellites after death of putative sire: case report, *Acta Vet. Hung.*, 49, 269–273, 2001.

38. Mellersh, C.S., Hitte, C., Richman, M., Vignaux, F., Priat, C., Jouquand, S. et al., An integrated linkage-radiation hybrid map of the canine genome, *Mammal. Genome*, 11, 120–130, 2000.

39. Hellmann, A.P., Rohleder, U., Eichmann, C., Pfeiffer, I., Parson, W., Schleenbecker, U.U., A Proposal for Standardization in Forensic Canine DNA Typing: Allele Nomenclature of Six Canine-Specific STR Loci, *J. Forens. Sci.*, 51(2): 274–281, 2006.

40. Parker, H.G., Kim, L.V., Sutter, N.B., Carlson, S., Lorentzen, T.D., Malek, T.B. et al., Genetic structure of the purebred domestic dog, *Science*, 304, 1160–1164, 2004.

41. Dolf, G., Schlapfer, J., Switonski, M., Stranzinger, G., Gaillard, C., and Schelling, C., The highly polymorphic canine microsatellite ZuBeCa4 is localized on canine chromosome 3q15-q18, *Anim. Genet.*, 29, 403–404, 1998.

42. Dolf, G., Schelling, C., Stahlberger-Saitbekova, N., Fu, B., Schlapfer, J., and Yang, F., Seven cosmid-derived canine microsatellites, *Anim. Genet.*, 31,411–412, 2000.

43. Francisco, L.V., Langston, A.A., Mellersh, C.S., Neal, C.L., and Ostrander, E.A., A class of highly polymorphic tetranucleotide repeats for canine genetic mapping, *Mamm. Genome*, 7, 359–362, 1996.

44. Holmes, N.G., Dickens, H.F., Parker, H.L., Binns, M.M., Mellersh, C.S., and Sampson, J., Eighteen canine microsatellites, *Anim. Genet.* 26, 132–133, 1995.

45. Holmes, N.G., Dickens, H.F., Neff, M.W., Mee, J.M., Sampson, I., and Binns, M.M., Nine canine microsatellites, *Anim. Genet.*, 29, 477, 1998.
46. Jonasdottir, T.J., Dolf, G., Sletten, M., Aarskaug, T., Schelling, C., Schlapfer, J. et al., Five new linkage groups in the canine linkage map, *Anim. Genet.* 30,366–370, 1999.
47. Ladon, D., Schelling, C., Dolf, G., Switonski, M., and Schlapfer, J., The highly polymorphic canine microsatellite ZuBeCa6 is localized on canine chromosome 5q12-q13, *Anim. Genet.*, 29, 466–467, 1998.
48. Neff, M.W., Broman, K.W., Mellersh, C.S., Ray, K., Acland, G.M., Aguirre, G.D. et al., A second-generation genetic linkage map of the domestic dog, *Canis familiaris, Genetics*, 151, 803–820, 1999.
49. Ostrander, E.A., Sprague, G.F., Jr., and Rine, J., Identification and characterization of dinucleotide repeat (CA)n markers for genetic mapping in dog, *Genomics*, 16, 207–213, 1993.
50. Richman, M., Mellersh, C.S., Andre, C., Galibert, F., and Ostrander, E.A., Characterization of a minimal screening set of 172 microsatellite markers for genome-wide screens of the canine genome, *J. Biochem. Biophys. Meth.*, 47(1–2), 137–149, 2001.
51. Shibuya, H., Collins, B.K., Huang, T.H., and Johnson, G.S., A polymorphic (AGGAAT)n tandem repeat in an intron of the canine von Willebrand factor gene, *Anim. Genet.*, 25, 122, 1994.
52. Shibuya, H., Collins, B.K., Collier, L.L., Huang, T.H., Nonneman, D., and Johnson, G.S., A polymorphic (GAAA)n microsatellite in a canine Wilms tumor 1 (WT1) gene intron, *Anim. Genet.*, 27(1), 59–60, 1996.
53. Tiret, L., Kessler, J.L., Bentolila, S., Faure, S., Bach, J.M., Weissenbach, J. et al., Assignation of highly polymorphic markers on a canine purebred pedigree, *Mamm Genome*, 11, 703–705, 2000.
54. Zajc, I. and Sampson, J., DNA microsatellites in domesticated dogs: application in paternity disputes, *Pflugers Arch.*, 431, R201–R202, 1996.
55. Zajc, I., Mellersh, C., Kelly, E., and Sampson, J., A new method of paternity testing for dogs, based on microsatellite sequences, *Vet. Rec.* 135, 545–547, 1994.
56. Zajc, I. and Sampson, J., Utility of canine microsatellites in revealing the relationships of pure bred dogs, *J. Hered.*, 90, 104–107, 1999.
57. The Canine Radiation Hybrid Project, 2003, www.recomgen.univ-rennes1.fr/doggy.html.
58. FHCRC Dog Genome Project, 2003, www.fhcrc.org/science/dog_genome. 54.
59. Balding, D.J. and Nichols, R.A., DNA profile match probability calculation: how to allow for population stratification, relatedness, database selection and single bands, *Forens. Sci. Int.*, 64, 125–140, 1994.
60. Foreman, L.A. and Evett, I.W., Statistical analyses to support forensic interpretation for a new ten-locus STR profiling system, *Int. J. Legal Med.*, 114, 147–155, 2001.

61. Waits, L.P., Luikart, G., and Taberlet, P., Estimating the probability of identity among genotypes in natural populations: cautions and guidelines, *Mol. Ecol.*, 10, 249–256, 2001.
62. Sutter, N.B., Eberle, M.A., Parker, H.G., Pullar, B.J., Kirkness, E.F., Kruglyak, L. et al., Extensive and breed-specific linkage disequilibrium in *Canis familiaris*, *Genome Res.*, 14, 2388–2396, 2004.
63. Taberlet, P., Waits, L.P., and Luikart, G., Noninvasive genetic sampling: look before you leap, *Trends Ecol. Evol.*, 14, 323–327, 1999.
64. Evett, I.W. and Weir, B.S., *Interpreting DNA Evidence: Statistical Genetics for Forensic Scientists*, Sinauer, Sanderland, MA, 1998.
65. DeNise, S., Johnston, E., Halverson, J., Marshall, K., Rosenfeld, D., McKenna, S. et al., Power of exclusion for parentage verification and probability of match for identity in American Kennel Club breeds using 17 canine microsatellite markers, *Anim. Genet.*, 35, 14–17, 2004.
66. Koskinen, M.T., Individual assignment using microsatellite DNA reveals unambiguous breed identification in the domestic dog, *Anim. Genet.*, 34, 297–301, 2003.
67. Angleby, H. and Savolainen, P., Forensic informativity of domestic dog mtDNA conrol region sequences, *Forens. Sci. Int.*, in press.
68. Overall, K.L. and Love, M., Dog bites to humans—demography, epidemiology, injury, and risk, *J. Am. Vet. Med. Assoc.*, 218, 1923–1934, 2001.
69. www.maulkorbzwang.de/statistiken.htm.
70. Ozanne-Smith, J., Ashby, K., and Stathakis. V.Z., Dog bite and injury prevention—analysis, critical review, and research agenda, *Inj. Prev.*, 7, 321–326, 2001.
71. Gershman, K.A., Sacks, J.J., and Wright, J.C., Which dogs bite? A case-control study of risk factors, *Pediatrics*, 93, 913–917, 1994.
72. Anon., Nonfatal dog bite-related injuries treated in hospital emergency departments—United States, 2001, *MMWR Morb. Mortal W. Rep.*, 52, 605–610, 2003.
73. Weiss, H.B., Friedman, D.I., and Coben, J.H., Incidence of dog bite injuries treated in emergency departments, *JAMA*, 279, 51–53, 1998.
74. Brogan, T.V., Bratton, S.L., Dowd, M.D., and Hegenbarth, M.A., Severe dog bites in children, *Pediatrics*, 96, 947–950, 1995.
75. Sacks, J.J., Sattin, R.W., and Bonzo, S.E., Dog bite-related fatalities from 1979 through 1988, *JAMA*, 262, 1489–1492, 1989.
76. Sacks, J.J., Kresnow, M., and Houston, B., Dog bites: how big a problem? *Inj. Prev.*, 2, 52–54, 1996.
77. Sacks, J.J., Lockwood, R., Hornreich, J., and Sattin, R.W., Fatal dog attacks, 1989–1994, *Pediatrics*, 97, 891–895, 1996.
78. Wiseman, N.E., Chochinov, H., and Fraser, V., Major dog attack injuries in children, *J. Pediatr. Surg.* 18, 533–536, 1983.

79. Sacks, J.J., Sinclair, L., Gilchrist, J., Golab, G.C., and Lockwood, R., Breeds of dogs involved in fatal human attacks in the United States between 1979 and 1998, *J. Am. Vet. Med. Assoc.*, 217, 836–840, 1983.

80. Sweet, D., Lorente, J.A., Valenzuela, A., Lorente, M., and Villanueva, E., PCR-based DNA typing of saliva stains recovered from human skin, *J. Forens. Sci.* 42, 447–451, 1983.

81. Sweet, D. and Hildebrand, D., Saliva from cheese bite yields DNA profile of burglar: a case report, *Int. J. Legal Med.* 112, 201–203, 1999.

82. Savolainen, P. and Lundeberg, J., Forensic evidence based on mtDNA from dog and wolf hairs, *J. Forens. Sci.* 44, 77–81, 1999.

83. Parson, W., Pegoraro, K., Niederstätter, H., Föger, M., and Steinlechner, M., Species identification by means of the cytochrome b gene, *Int. J. Legal Med.*, 114, 23–28, 2000.

84. Savolainen, P., Rosen, B., Holmberg, A., Leitner, T., Uhlen, M., and Lundeberg, J., Sequence analysis of domestic dog mitochondrial DNA for forensic use, *J. Forens. Sci.* 42, 593–600, 1997.

85. Savolainen, P., Arvestad, L., and Lundeberg, J., A novel method for forensic DNA investigations: repeat-type sequence analysis of tandemly repeated mtDNA in domestic dogs, *J. Forens. Sci.*, 45, 990–999, 2000.

86. Schneider, P.M., Seo, Y., and Rittner, C., Forensic mtDNA hair analysis excludes a dog from having caused a traffic accident, *Int. J. Legal Med.*, 112, 315–316, 1999.

87. Wetton, J.H., Higgs, J.E., Spriggs, A.C., Roney, C.A., Tsang, C.S., and Foster, A.P., Mitochondrial profiling of dog hairs, *Forens. Sci. Int.* 133, 235–241, 2003.

88. Kim, K.S., Lee, S.E., Jeong, H.W., Ha, J.H., The complete nucleotide sequence of the domestic dog (*Canis familiaris*) mitochondrial genome, *Mol. Phylogenet. Evol.*, 10, 210–220, 1998.

89. Pereira, L., Van Asch, B., and Amorim, A., Standardisation of nomenclature for dog mtDNA D-loop: a prerequisite for launching a *Canis familiaris* database, *Forens. Sci. Int.*, 141, 99–108, 2004.

5

STR-Based Forensic Analysis of Felid Samples from Domestic and Exotic Cats

MARILYN A. MENOTTI-RAYMOND, PH.D., VICTOR A. DAVID, M.S., AND STEPHEN J. O'BRIEN, PH.D.

Contents

5.1 Introduction

Humankind has long had a fascination with the cat. Just how long is suggested by the presence of skeletal remains of *Felis sylvestris*, the wild cat from which the house cat was domesticated, closely associated with human remains in a grave site in ancient ruins in Cyprus, dating to 10,000 years ago.[1] The wild cat was not native to the island, and the presence of the cat with human remains suggests that the cat was part of the human entourage. Cats may have served as mousers for early humans, helping to control rodent infestation of food stocks.[2,3] Cats have been found as mummified relics in burial vaults, a sign that the human's reverence of cats approached near deification in early Neolithic societies of the Middle East. As shipboard companions, cats ultimately were distributed worldwide along the major trade routes.Today, some 75 million cats reside in households in the United States alone.[4] D'Andrea has demonstrated that it is virtually impossible to enter the home where a cat or dog resides without becoming a carrier of its hair.[5] With a cat residing in approximately one in three households, it is not surprising that cat hairs are often part of the physical evidence associated with crime scenes. Hairs from a pet can be indicative of a perpetrator's presence at a crime scene or provide evidence of a connection between victim and perpetrator.

5.2 History of Cat Hair Analyses

Early attempts at forensic analysis of cat hairs relied on morphological criteria,[6,7] and identification was made on a species or, at best, breed level. On the molecular level, the first attempt at characterization of hairs to identify species utilized isoelectric focusing of keratins.[8] Jeffreys, Wilson, and Thein[9] first demonstrated the potential of genetic individualization of human samples on the DNA level, with the characterization of a high level of genetic variation at highly repetitive minisatellite loci. DNA fingerprints were used in the first identification of human DNA in an immigration test case.[10] Hybridization of human minisatellite probes to restriction endonuclease digests of cat and dog DNA also generated multilocus DNA fingerprints, which offered the possibility of genetic individualization of human's companion animals.[11]

The first cat-specific multilocus minisatellite probes developed by Gilbert et al.[12] were used to assess the relatedness of members in lion prides and cheetah populations[13] and to infer social dynamics of lion pride structure.[12] Multilocus probes became the forerunner of site-specific single locus minisatellite probes, or VNTR (variable number of tandem repeats). These loci had the advantage in that they could be well characterized on a molecular level, including genetic map location and mutation rate. Additionally, population genetic databases could be generated with which to compute match probabilities. Single-locus VNTR probes had not yet been characterized in other species before the report of simple tandem repeat (STR) markers,[14,15] which had clear advantages over VNTRs.

STR loci were short in length, generally having fewer than 100 base pairs (bp), were abundant in all eukaryotic genomes examined, and were highly polymorphic.[14] With the advent of rapid development in molecular genetic technologies to aid in the sequencing of the human genome,[16,17] STRs offered the opportunity to genotype multiple highly polymorphic unlinked loci in a single PCR reaction, using less than a nanogram of genomic DNA.[18,19] Human forensics was revolutionized, with a rather rapid shift from VNTR analysis to STR profiling.[20,21] With the routine use of STR loci applied to the genetic individualization of human samples, came the realization of the potential of samples of nonhuman origin. Genetic linkage maps incorporating STRs with coding loci were rapidly developed in companion and commercial animals for the mapping and characterization of genes of health-related and commercial interest,[22–29] opening the potential of STR application to forensic applications.

At the National Cancer Institute's Laboratory of Genomic Diversity, we have maintained an interest in the domestic cat as a model of hereditary and infectious disease with which to improve human health.[30] Over 263 hereditary disease pathologies have been reported in the cat,[31] many of which bear homology to human hereditary disease. In an effort to identify genes associated with disease pathology in the cat, we have developed genetic linkage and radiation hybrid maps of the domestic cat incorporating both coding genes and STR loci.[29,32–35]

5.3 Case Examples

The characterization of hundreds of feline STR loci has provided a powerful genetic toolkit for genetic analysis of both domestic cats and related felids. The STR loci we have developed in the domestic cat have proven useful in conservation genetic studies involving a number of endangered felids, including the lynx, lion, jaguar, tiger, cheetah, mountain lion, leopard, Geoffrey's cat, tigrina, pampas cat, and wild cat.[36–43] Additionally, STR genotyping has been used for parentage studies and genetic individualization in exotic felids (Warren Johnson, pers. comm.).[43–45] Our first application of genetic individualization of felids using domestic cat STR loci was performed with samples from mountain lions at the request of state wildlife officials. In 1995 we were able to determine that a mountain lion captured in northern Florida was not an endangered Florida Panther, but rather an offspring of two mountain lions from Texas. The parents had been released into the area by the Florida Game and Freshwater Fish Commission as a part of a feasibility study for mountain lion reintroduction.[44] It was important to ascertain the origin of this animal, as had it been a true Florida Panther, both the animal and the area in which it resided could have been afforded special protection under the Endangered Species Act.

A second application of STR loci to the genetic individualization of felid samples arose from a mountain lion attack in California. Early on the morning of December 4, 1994, a woman hiking alone was attacked and killed by a mountain lion in the Cuyamacha Rancho State Park near San Diego.

California Department of Fish and Game officials killed a mountain lion in the area and submitted samples for DNA analysis in order to confirm that they had killed the animal responsible for the attack. STR profiles for eight loci, generated from DNA isolated from mountain lion hairs found on the victim and wound swabs, matched profiles obtained from the mountain lion shot by state officials (Melznie Culver, personal communication).

The first application of feline STR loci to genetic individualization of domestic cats came in response to an inquiry from the Royal Canadian Mounted Police (RCMP). Could we generate a DNA fingerprint from cat hairs found at a crime scene and compare them to those of a suspect's pet cat? The case concerned a 32-year-old woman who had disappeared from her home in Richmond, Prince Edward Island, Canada, on October 3, 1994. Her abandoned car, discovered within a few days in a wooded setting not far from her home, was stained with blood determined to be that of the missing woman. Three weeks later, a search team discovered a bag containing a man's leather jacket and tennis shoes, stained with the victim's blood, 8 kilometers from the victim's home. White hairs found inside the lining of the jacket were identified by the RCMP's Halifax forensics laboratory (A. E. Evers, personal communication) to be cat hairs. The prime suspect, the estranged common-law husband of the victim, lived at the home of his parents, with a white cat named Snowball (Figure 1).

We agreed to perform the analyses. Higuchi had previously demonstrated that nuclear amplification could be performed with DNA extracted from a single human hair root using the polymerase chain reaction,[46] while

Figure 1 Snowball, the suspect's pet cat. (Stephen J. O'Brien, Teers of the Cheetah, St. Martin's Press, New York, 2003.)

Morin had reported on genetic individualization of single-hair specimens of free-ranging chimpanzees.[47] The critical question with the evidentiary hairs was whether any single hair would have enough DNA to perform the analyses. How old were the hairs? Did they have roots? How much DNA is there in a cat hair root? Preliminary experiments in our laboratory indicated that while DNA recovery was typically 10-fold lower from cat hairs than from human hairs, sufficient DNA for STR typing could be obtained from a single hair.[48]

Constable Roger Savoie from the RCMP arrived in Washington, DC with two sealed evidence containers; one with 27 white hairs from the jacket lining and another with blood drawn from the subpoenaed cat. Following microscopic examination of the hairs, four candidate specimens with visible roots were selected for analysis. The hairs were cut into two fragments, a root portion of approximately 10 millimeters and the remaining hair shaft. Hair fragments were washed and DNA was extracted following standard proteinase K digestion and phenol-chloroform extraction methods.[49] Agarose gel electrophoresis of approximately 20% of each hair fragment extraction product demonstrated a high molecular weight band for one of the hair roots, estimated at approximately 17 nanograms (ng) (Figure 2). Since we had no prior knowledge of the age or condition of the hairs, we were concerned about DNA recovery. It came as a great relief, when we flipped the switch on the ultraviolet light box to view our agarose gel, that a DNA band was visible for one of the cat root extraction products.

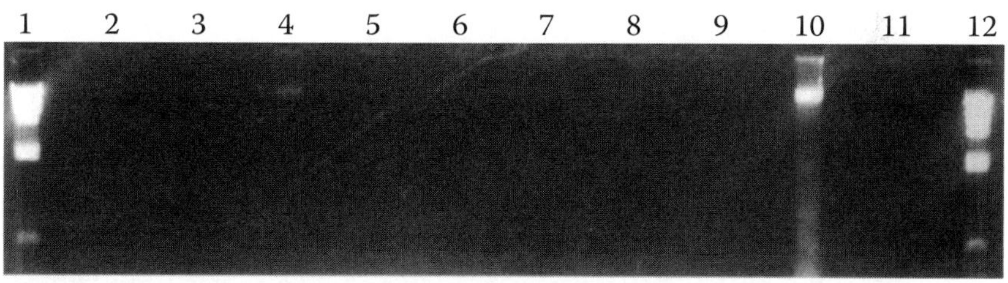

Figure 2 Agarose gel demonstrating recovery of high-molecular-weight DNA from cat hair found in jacket of presumed murderer. Four evidentiary cat hairs and one control human hair were cut into root and shaft fragments that were processed separately for DNA isolation. Lane 1—size standard, lane 2—cat hair 1 root, lane 3—cat hair 1 shaft, lane 4—cat hair 2 root, lane 5—cat hair 2 shaft, lane 6—cat hair 3 root, lane 7—cat hair 3 shaft, lane 8—cat hair 4 root, lane 9—cat hair 4 shaft, lane 10—human hair root, lane 11—human hair shaft, lane 12—size standard. Cat hairs 1 and 2 were untreated prior to digestion, however hairs 3 and 4 had been mounted in permount by the Royal Canadian Mounted Police for microscopic evaluation. Note the presence of high-molecular-weight DNA recovered from cat hair root 2 and human hair root.

Extraction products from all roots and shafts and reagent controls were amplified singly with 10 dinucleotide STR loci that had been selected for the analysis[46] based on robustness, absence of linkage, and low "stutter."[50] PCR products were generated from DNA extracted from the one root with visible DNA.[49] Subsequently, DNA was isolated from the blood sample, and PCR products from the hair root and blood extractions were electrophoresed in the same gel. PCR products were fluorescently labeled and electrophoresed in 6% denaturing polyacrylamide gels using an Applied Biosystems 373A Automated DNA Sequencer. Figure 3 demonstrates electropherograms of

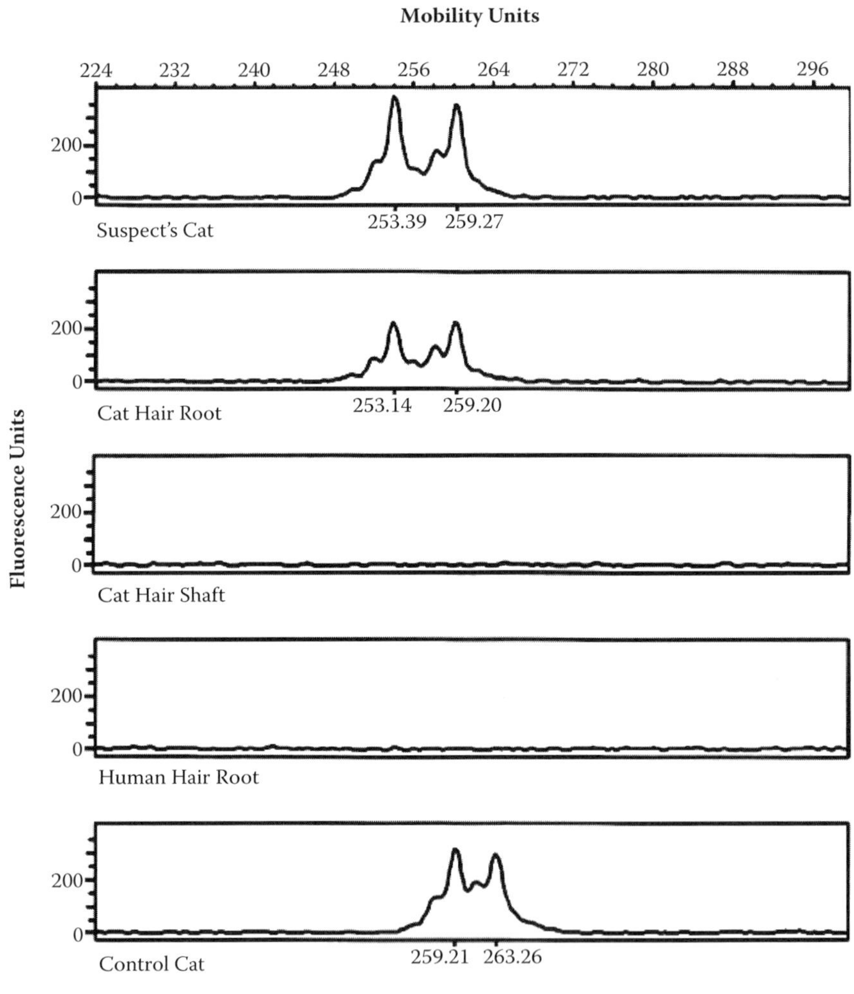

Figure 3 Electropherograms resulting from amplification of STR locus FCA 080 with evidentiary and control cat DNA samples. The alleles obtained from the cat hair root were judged to match the alleles from the suspect's cat for this locus.

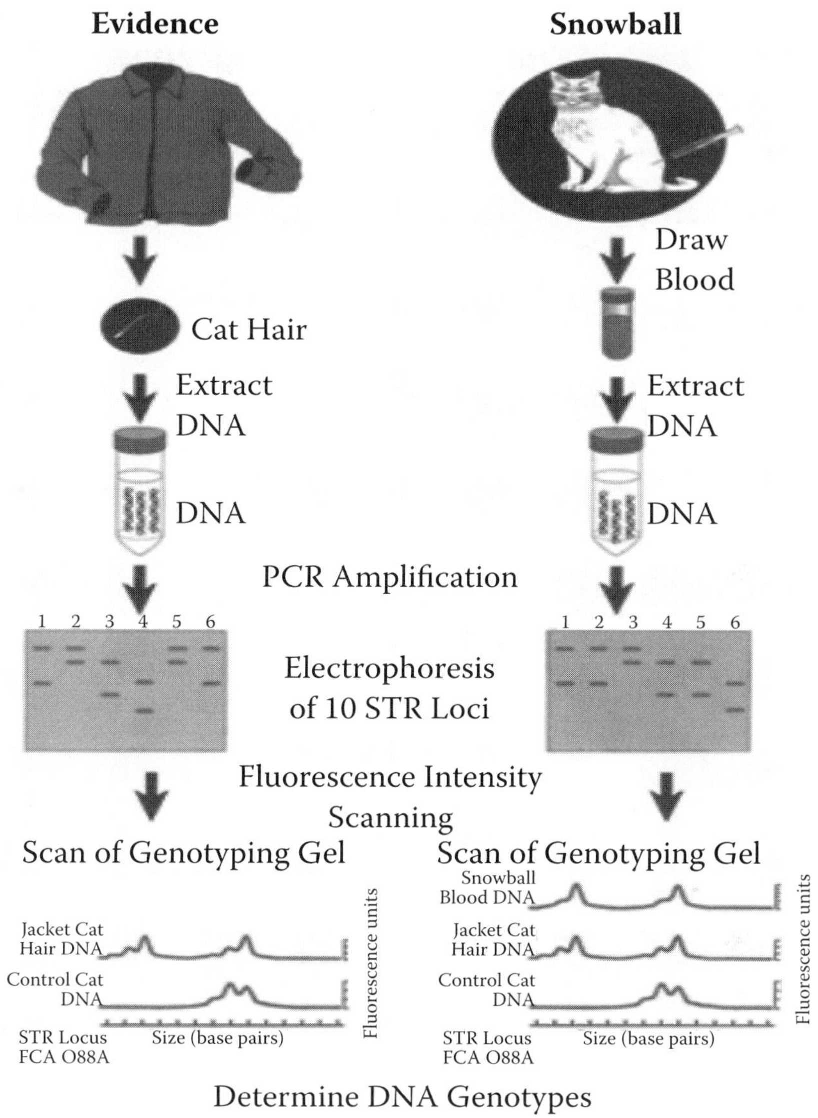

Figure 4 Graphics presented to the Prince Edward Island jury depicting the sequence of steps in generating DNA profiles for evidentiary hairs and blood sample from Snowball. (**See color insert after p. 108.**)

PCR products generated for one locus. Figure 4 illustrates the sequence of steps in the analyses of hair and blood specimens, which was presented to the Prince Edward Island jury.

Composite STR genotypes amplified from hair root and blood DNA were determined to match at all 10 loci (seven heterozygous and three

Comparison of Snowball and Jacket Cat Hair STR Genotypes

STR Locus	DNA Source	Allele	Allele Size (bin)	Size Difference (bp) |Snowball Blood - Hair|		STR Locus Match Window (bp)	Conclusion
FCA 026	Snowball	1	147.83				
	Jacket cat hair	1	148.11	0.28	<	0.37	MATCH
	Snowball	2	143.73				
	Jacket cat hair	2	143.73	0	<	0.37	MATCH
FCA 043	Snowball	1	126.38				
	Jacket cat hair	1	126.29	0.09	<	0.59	MATCH
	Snowball	2	120.52				
	Jacket cat hair	2	120.50	0.02	<	0.59	MATCH
FCA 080	Snowball	1	259.27				
	Jacket cat hair	1	259.20	0.07	<	0.30	MATCH
	Snowball	2	253.39				
	Jacket cat hair	2	253.14	0.25	<	0.30	MATCH
FCA 088A	Snowball	1	121.91				
	Jacket cat hair	1	121.91	0	<	0.42	MATCH
	Snowball	2	110.50				
	Jacket cat hair	2	110.50	0	<	0.42	MATCH
FCA 126	Snowball	1	143.41				
	Jacket cat hair	1	143.41	0	<	0.53	MATCH
	Snowball	2	141.08				
	Jacket cat hair	2	141.08	0	<	0.53	MATCH
FCA 132	Snowball	1	152.73				
	Jacket cat hair	1	152.73	0	<	0.27	MATCH
	Snowball	2	150.69				
	Jacket cat hair	2	150.69	0	<	0.27	MATCH
FCA 149	Snowball	1	132.02				
	Jacket cat hair	1	131.80	0.22	<	0.29	MATCH
	Snowball	2	128.05				
	Jacket cat hair	2	128.07	0.02	<	0.29	MATCH
FCA 058	Snowball	1	229.03				
	Jacket cat hair	1	229.24	0.21	<	0.36	MATCH
FCA 090	Snowball	1	93.54				
	Jacket cat hair	1	93.54	0	<	0.46	MATCH
FCA 096	Snowball	1	210.95				
	Jacket cat hair	1	211.00	0.05	<	0.25	MATCH

Figure 5 Graphics presented to the Prince Edward Island jury presenting matching composite STR profiles generated from DNA extracted from an evidentiary cat hair and the blood of Snowball.

homozygous) STR loci (Figure 5). We used guidelines as developed by the RCMP to determine whether we had matching profiles.[51] In order to establish criteria for declaring a match between STR alleles of similar size, human forensic laboratories empirically determine a size difference threshold ("match window"), that defines acceptable variation in migration between any two measured alleles to conclude that they match.[51] The match window provides an empirical determination of precision for each STR locus from a sample of multiple ascertainments of identical alleles. The magnitude of the match window is based on the difference observed in the largest and smallest size estimation for identical alleles, which were electrophoresed on the same gel.

5.4 Determination of a DNA Match

We determined empirically the match window for the 10 feline STR loci by comparing migration (size) estimates of alleles identical by descent in a 70-member feline pedigree used to generate a genetic recombination map of the cat.[29] The difference in size estimation between multiple individuals for each allele was determined. All size comparisons were made of multiple cat DNA samples electrophoresed on the same gel. The maximum size difference observed after examination of all allele size classes determined the level of precision or match window for that particular STR locus.[48] Alleles that differed in size by a quantity less than or equal to the match window were judged to be a match.[49] Using this set of criteria, all pairs of alleles for the 10 loci (17 pair-wise comparison) were determined to be a match (Figure 5).[49]

The likelihood of a match between the hair genotype and a random individual was estimated from the frequency of the composite genotype in the population at large using allele frequency estimates from two STR population surveys: 19 unrelated cats from Prince Edward Island and nine cats from around the United States.[49] Although small, the island sample was adequate (95% confidence) to detect any STR allele present at a frequency of 9.5% or higher.[49] The two populations showed appreciable allelic variation and remarkable population genetic similarity (as opposed to geographic population substructure). The incidence of the composite hair genotype for the seven heterozygous loci, estimated using the product rule[52] and minimum allele frequency estimates for rare alleles,[53] was 2.2×10^{-8} and 6.9×10^{-7} for the island and the United States population databases, respectively.[49]

The results of our analysis were presented and admitted to the Supreme Court of Prince Edward Island. The jury convicted the defendant of second-degree murder on July 19, 1996. Matching composite profiles for cat hair and blood DNA were part of the evidence introduced into court, including human DNA evidence, which was taken into consideration by the jury. This legal precedent for introducing animal genetic individualization in a homicide trial stimulated interest from a number of forensic laboratories for feline testing.

5.5 Development of an STR Typing System for Cats

Under support from the National Institutes of Justice (NIJ), we have continued our research efforts toward the generation of a formalized STR forensic typing system for the genetic individualization of domestic cat specimens. Our goals included:

1. Isolation, characterization, and mapping of candidate STR tetranucleotide repeat loci for a forensic panel
2. Selection of a forensic panel of 10 STR loci and a gender identifying sequence tagged site (STS)
3. Development of a multiplex amplification protocol
4. Development of a method to quantify DNA yield from single-hair isolates to evaluate whether STR analysis is feasible
5. Generation of a population genetic database using the STR forensic panel
   a. Collection of samples (blood/buccal swab) representing major recognized cat breeds in the United States
   b. Genotyping and analysis of the forensic STR panel in collected samples

5.6 Selection of the CAT STR Marker Panel

Forensic analysis of human specimens utilizes tetranucleotide STRs, loci that exhibit the tandem repeat of a four-base-pair motif. Tetranucleotide STRs minimize the generation of "stutter band" products generated during PCR amplification,[50] which can complicate the interpretation of genotypes from mixed DNA samples. In order to increase the number of candidate tetranucleotide STR loci for a domestic cat forensic panel, a set of 49 tri- and tetranucleotide STR loci were isolated from STR-enriched genomic libraries.[33,54] The markers were incorporated into genetic maps of the domestic cat relative to 579 coding genes and 255 STR loci[33] in order to select markers that were unlinked. Screening of the 49 loci in a small panel of outbred domestic cats identified a panel of 22 loci, which demonstrated high variability or heterozygosity, as candidates for a forensic panel.[54] The 22 loci were genotyped in a sample set of 28 cat breeds (three to 10 animals/breed, n = 213), in order to select a panel of markers with the highest discriminating power for forensic analysis.

As a result of these analyses, a set of 11 highly polymorphic loci was selected as a domestic cat forensic typing panel. Selection criteria were based on identifying loci that were genetically unlinked and demonstrated: 1) high heterozygosity across multiple cat breeds, 2) an absence of cross-species amplification, and 3) a single target of amplification. The loci are well distributed across the cat's 18 autosomes, with seven loci mapping to independent chromosomes and four loci located on separate arms of the largest cat chromosomes.[33]

5.7 PCR Amplification Conditions for the Cat Multiplex

An amplification protocol has been developed for the coamplification of the 11 STR loci.[54,55] Additionally, a fragment of the SRY gene on the cat Y-chromosome is amplified in the multiplex in order to identify the gender of the sample.[54,55] The PCR products for the 11-plex STR amplification, labeled with one of four fluorescent tags, were designed to fall in a size range from 100 to 415 base pairs, with the SRY product detectable at 96 base pairs (Figure 6). Sequences for the multiplex primer pairs, the final concentration of primers determined empirically to generate a balanced product profile, and amplification conditions are presented in *The Journal of Forensic Science*.[54]

The multiplex is amplified with the same thermal cycling conditions used in commercial STR kits for the genotyping of human DNA, namely 28 cycles of PCR with an annealing temperature of 59°C. The use of common amplification conditions and PCR setup and performance should assist in easing the adoption of the cat multiplex by forensic DNA laboratories already performing human STR typing. The 12-member assay is robust and generates a balanced product profile (Figure 7). The cat multiplex is typically generated with 1–5 nanograms of genomic DNA for population samples. Initial sensitivity assays demonstrate full product profiles can be generated from 125 picograms of genomic DNA, with an absence of "allele dropout," though more in-depth sensitivity studies are currently in progress.[56]

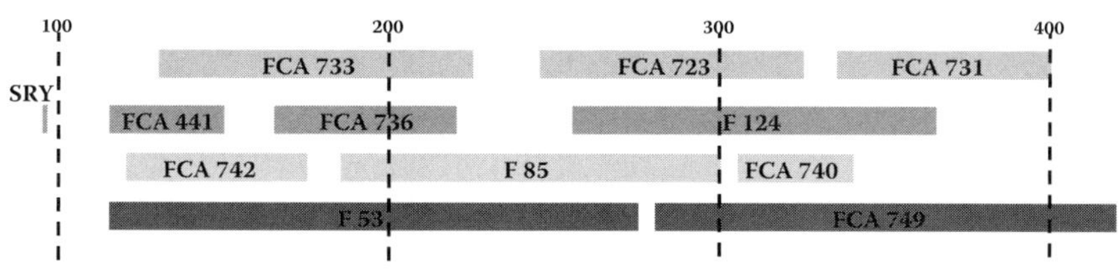

Figure 6 Fluorescent dyes and size ranges for domestic cat 12-plex as observed in a 1,040 member domestic cat genetic database generated from 38 cat breeds.[54] **(See color insert after p. 108.)**

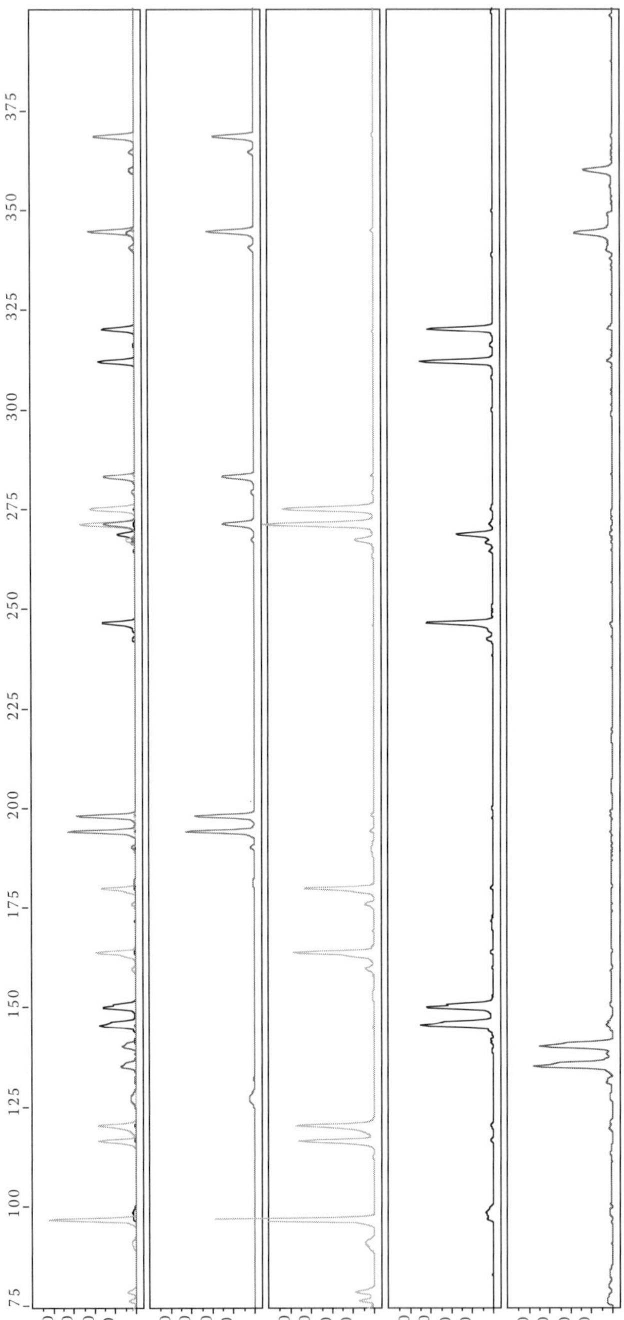

Figure 7 (**See color insert after p. 108.**) Electrophoretogram of PCR products of 12-member multiplex amplified from 4 ng of male genomic DNA (upper panel). Lower panels demonstrate PCR products labeled with fluorescent tags FAM (3 STR), VIC (3 STR, SRY gene), NED (3 STR), and PET (2 STR).

5.8 Cat Reference Population Databases

We generated a genetic database for the domestic cat in recognized cat breeds, as factors associated with the generation and propagation of breeds (founder effects, small effective population sizes, the use of popular sires, inbreeding, artificial selection) could have a major impact on genetic profiles in breeds and the generation of population substructure. Although the majority of domestic cats maintained as pets in the United States are mixed breed in nature (approximately 97%), an assumption cannot be made *a priori* that an evidentiary sample came from a mixed breed animal. An STR panel developed for forensic analysis of cat samples must have adequate resolution for genetic individualization within the reduced gene pools of cat breeds. The earliest fossil records linked to the domestication of the cat date to approximately 9,500 years ago from Cyprus.[1] What is known about the history of cat breeds is generally anecdotal in nature,[57,58] but all breeds are relatively recent in origin from an evolutionary standpoint, within hundreds of years. The majority of breeds recognized in the United States by the two largest cat registries (Cat Fanciers' Association [CFA] [http://www.cfainc.org/] and The International Cat Association [TICA] [http://www.tica.org/]) have received breed recognition only within the last 100 years. These breeds are recent phylogenetic lineages that capture different combinations of coat color, hair length, patterning, and distinctive morphological traits reflecting different combinations at likely fewer than 30 loci, many with homologous counterparts in coat color genes of mouse and other domestic species.

The panel of 11 felid-specific STR loci has been used to generate a population genetic database of the major domestic cat breeds recognized in the United States today by the CFA and TICA with which to compute composite match probabilities. Blood and cheek swab samples were collected from cats registered with either the CFA or TICA through mailings or contact with owners at cat shows. The sample set consisted of 1,040 individuals of 38 recognized cat breeds. A small sample set of outbred domestic cats ($n = 24$) was included for comparison to the cat breeds.

The heterozygosity for the 11-locus set observed across cat breeds is presented in Figure 8. The loci demonstrate moderate locus heterozygosities within cat breeds with an 11-locus average heterozygosity of 0.71, ranging from 0.56 (locus FCA 740) to 0.78 (locus FCA 723). A high 11-locus composite heterozygosity of 0.85 was observed in outbred domestic cats. The 38 cat breeds demonstrate moderate to high composite locus heterozygosities for the 11-plex, from 0.55 observed in the Havana Brown breed to 0.84 observed in the Norwegian Forest Cat.[54] Cat breeds demonstrated considerable variation in locus-specific heterozygosities, allelic size ranges and distributions, and allele frequencies.[59] Given the locus heterozygosities, the

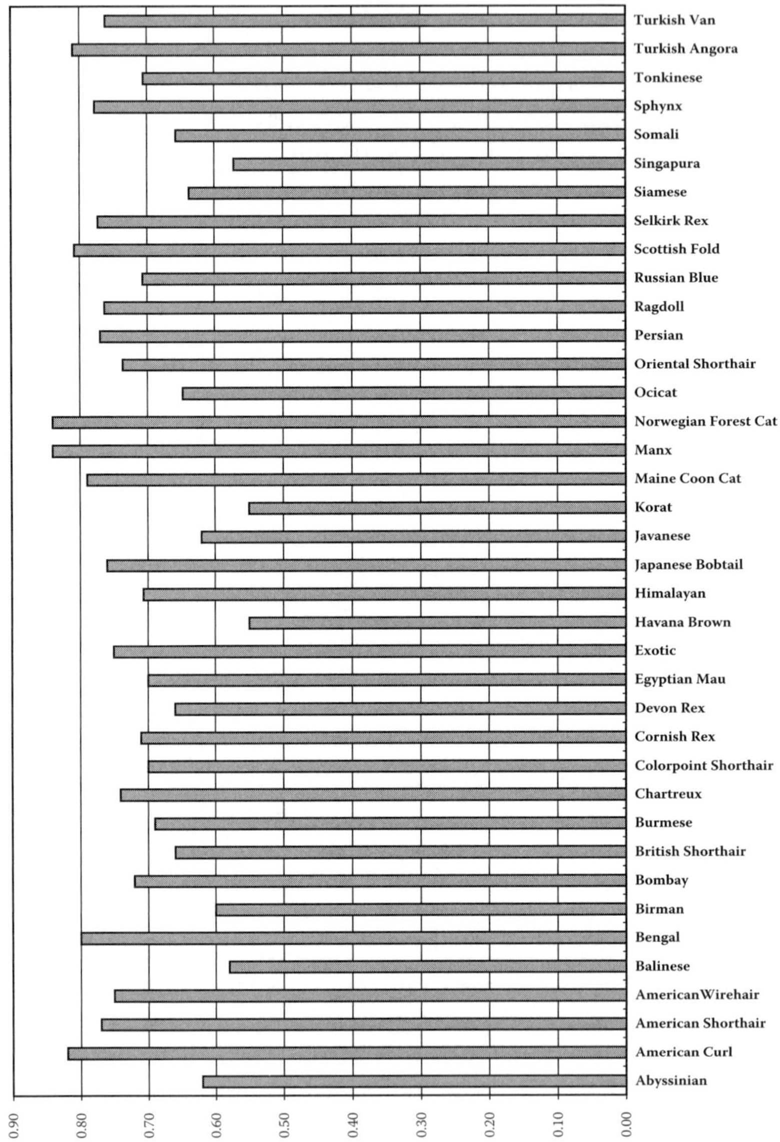

Figure 8 Average locus heterozygosities observed in 38 cat breeds for the 11-member STR multiplex. The total sample collection was 1,040 animals, with an average of 37 animals/breed. Average heterozygosity observed for a small sample of outbred cats was 0.85.

power for genetic individualization of domestic cat samples using the multiplex is high in cat breeds, ($P_m = 2.4 \times 10^{-6}$ to 6.4×10^{-13}), and high for outbred domestic cats, as demonstrated by a P_m of 7.8×10^{-13}. The 11-member average heterozygosity of 0.85 and P_m observed for mixed breed cats (Table 1) suggest that the STR panel will have good potential for genetic discrimination across the more genetically diverse population of mixed-breed cats. Initial validation studies have been performed on the cat multiplex following the DNA Advisory Board's Recommendations for Quality Assurance.[56,60]

5.9 Determination of Species Specificity and Other Validation Studies

To examine species specificity of the multiplex, the 11 tetranucleotide STR loci were examined in a range of North American mammalian species, including dog, deer, rabbit, guinea pig, hamster, mouse, horse, cow, pig, ferret, mink, sheep, goat, brown bear, fox, badger, wolf, human, beaver, otter, raccoon, possum, skunk, mole, coyote, chipmunk, puma, ocelot, domestic cat, and two prokaryotes, *Sacchromyces cervesiae* and *Escherichia coli*. The multiplex displayed a high degree of specificity for DNA in the felid family with PCR products observed in ocelot, puma, and domestic cat and PCR products for two loci generated from brown bear DNA, another member of the Carnivore order.[56] Using the standard amplification conditions, no products were observed in any other mammalian species or the prokaryotes.

Under conditions of limiting DNA template, stochastic effects are often observed, in which a single allele of a heterozygous individual is amplified.[61] In order to determine the sensitivity of the multiplex, amplification was performed in a dilution series of domestic cat genomic DNA. With a minimum threshold of 50 relative fluorescence units (RFUs), a quantity of 0.125 ng was required in the two sample DNAs to detect both alleles of heterozygous loci in the 11-STR loci.[56] Genotyping of human DNA is performed with commercially available STR multiplex kits with DNA quantities as low as 0.125 nanograms[62] before allele dropout is observed. To examine reproducibility of the multiplex, PCR amplification profiles were generated from DNA extracted from blood and buccal samples from 13 domestic cats. Identical profiles were obtained for the duplicate samples from all individuals.[56]

Hair specimens are likely to be the most common type of sample from domestic cats associated with crime scenes. Genetic individualization of animal hair specimens is increasingly being employed in forensic cases.[44,63–65] A limiting factor in animal STR genetic profiling, particularly from hair specimens, will be the quantity of genomic DNA available for analysis. Cat hair roots are generally a poor source of genomic DNA. In initial analyses, we have obtained 30 nanograms of DNA from the very best fresh plucked guard hair roots,[66] which is on an order of 10 to 30 times less DNA than is available from a human hair root. Cat hairs that are aged, shed, or originate from the undercoat (fine wool hairs) prove to be a much poorer source of DNA.[66] Given the low yield of DNA from cat hair specimens, a goal of our feline STR forensic typing system has been to develop an assay to quantify DNA in order to determine the likelihood of STR genotyping success without compromising product available for alternate analyses, such as mitochondrial DNA typing, which requires less DNA.[67]

We have designed a method to estimate feline genomic DNA yield through a quantitative, or real-time PCR based assay.[68] Quantitative PCR assays measure the amount of PCR product at the completion of each cycle. Thus, by comparing the PCR product profile generated from a DNA source of unknown quantity with the product profiles of a DNA dilution standard, it is possible to estimate the amount of DNA in the unknown.[68,69] We have developed a highly sensitive quantitative PCR-based assay that targets SINE (small interspersed nuclear elements), a highly abundant repetitive element comprising approximately 10% of the cat nuclear genome.[70] SINE elements are also abundant in other mammalian genomes,[71–73] and are used to quantify human genomic DNA.[74]

Primer pairs were designed to amplify cat SINE elements from conserved regions in an alignment of 100 SINE elements sequenced in the domestic cat.[64] The assay monitors PCR product accumulation through a fluorescent dye, which binds to double-stranded DNA.[69] A standard curve of product accumulation generated from a dilution series of known DNA concentration is used to interpolate the concentration of the unknown. The assay is highly sensitive, can be performed rapidly using trace amounts of DNA, and detects feline genomic DNA at a concentration of 10 femtograms (fg) in a 20-μL reaction, following 30 cycles of amplification.[68] As primer pairs were designed in regions that demonstrate a high degree of sequence conservation across species, the assay is not felid specific.[68] However, we do not think that the lack of species specificity deters from the utility of the assay, as DNA mixtures are unlikely to be an issue with feline samples.

5.10 Mitochondrial DNA Analysis in Cats

In the last several years, mitochondrial (mt) DNA analysis has become a valuable resource for forensic analysis of samples considered inappropriate for STR profiling,[67,75] due to inadequate amounts of nuclear DNA or high degradation of the sample. MtDNA profiling of human hair samples, particularly samples consisting of a hair shaft alone, has been highly successful, and is additionally used for the most sensitive assays of degraded or ancient samples.[71,79] A significant proportion of cat hairs that are inappropriate for

STR profiling will be candidates for mtDNA genotyping. The entire nucleotide sequence of the domestic cat mtDNA genome has been published from our laboratory.[80] The structure and gene content of the cat mtDNA genome largely resembles the mtDNA of placental mammals.[80] Of import, domestic cat evolution is marked by the transposition of approximately 50% of the mitochondrial genome into the nuclear genome in an ancestor of the domestic cat, approximately 2 million years ago.[81] The 7.9 kilobase region, termed *Numt,* extends from the 3' end of the control region through 80% of the COII gene, and exhibits high sequence similarity with the homologous cytoplasmic mtDNA region (5.1% sequence divergence).[81] Forensic analyses of polymorphic sites in cat mtDNA will necessarily need to exclude the *Numt* region as coamplification of the cytoplasmic and nuclear mitochondrial DNA would generate a high degree of heteroplasmy, which would be difficult or impossible to interpret.

A study of variation within the mtDNA Control Region (CR) region of domestic cats conducted by Fridez[82] demonstrated the discriminating power of the mitochondrial CR in outbred domestic cats. Twenty-one polymorphic sites identified 14 haplotypes in a survey of 50 cats,[82] which included 42 outbred European short hairs, five Persian, two Siamese, and one Abyssinian.[82] No breed-specific haplotypes were exhibited. The power of discrimination of the 21 polymorphic sites was reported as 0.84.[82] Enormous forensic potential exists for mtDNA profiling of animal specimens with the development of expanded species specific mtDNA typing systems and databases.

5.11 Future Areas of Research for Characterization of Cat Samples

An additional source of forensic profiling from "nonhair" cat specimens that might be important in an investigation is the determination of coat color phenotype. In cases where nonhair specimens, such as blood, urine, feces, saliva, or bone are available, investigators might determine the coat color of the cat from which the samples originate. Mutations have been characterized in the domestic cat that are responsible for generation of melanistic or black hair color,[83] two color variants of the brown locus (chocolate and cinnamon),[84] and dilute hair color.[85] Additionally three separate mutations at the albino locus are responsible for progressive lack of melanin pigment deposition, that results in three identified coat color phenotypes: albino, Siamese, and Burmese.[86–88] Mutations for all of the coat color variants mentioned can easily be characterized with site-directed PCR and sequence analysis.[83–85,87,88]

The forensic analysis of biological samples has been revolutionized over the past 20 years, a metamorphosis from the analysis of the protein products of DNA to DNA itself. With the development of highly sophisticated methodologies and instrumentation, DNA analyses can generate match probabilities of a magnitude that can be considered unique.[89] Animal forensics is still in its infancy. However, we do not need to reinvent the wheel completely. Many of the practices and policies developed in human forensics can be adopted for treatment of animal specimens. An STR kit is currently under development for cat profiling, and a population genetic database of 1,040 individuals representing 38 breeds has been generated.[59] What is now required includes the development of commercial kits (for cats), allelic ladders for both cat and dog STR kits, validation studies, and the development or extensive population genetic databases that have been peer reviewed. Biological evidence from animals will play an increasing role in forensic casework as more tests are developed that can be easily implemented in the laboratory. There have already been a number of high-profile cases in which animal evidence has played a key role for the prosecution.[49,90–92] Additionally, an *ad hoc* survey of detectives across the country by the National Institute of Justice's (NIJ) Office of Science and Technology revealed a strong interest in using such evidence if it was available.

Acknowledgments

The authors wish to thank the National Institute of Justice for funding this research through interagency agreements to the National Cancer Institute's Laboratory of Genomic Diversity. We are especially thankful to John Butler (National Institute of Standards and Technology) for developing and optimizing a multiplex amplification protocol and for many helpful suggestions throughout this research. We gratefully acknowledge the Cat Fanciers' Association (CFA) and the International Cat Association (TICA) for their considerable help in facilitating sample collection from cat breeds and the hundreds of independent cat breeders who provided us with blood and buccal swab samples of their cats. We wish to thank Ellen Frazier (Publications Department, SAIC-Frederick, Inc., Frederick, MD) for all of the graphics illustrations. Leslie Wachter (SAIC-Frederick), Amy Snyder (SAIC-Frederick), and Nikia Coomber (NCI, LGD) were dedicated technicians in this research project. The content of this publication does not necessarily reflect the views or policies of the Department of Health and Human Services, nor does mention of trade names imply endorsement by the U.S. Government.

References

1. Vigne, J.D. et al., Early taming of the cat in Cyprus, *Science,* 304, 259, 2004.
2. Clutton-Black, J. 1999. *A natural history of domesticated mammals.* Cambridge: Cambridge University Press.

3. Malek, J., *The Cat in Ancient Egypt*, British Museum Press, London, 1993.
4. Pet Food Institute. http://Petfoodinstitute.org (Home page; accessed April 1, 2007).
5. D'Andrea, F., Fridez, F., and Coqouz, R., Preliminary experiments on the transfer of animal hair during simulated criminal behavior, *J. Forens. Sci.*, 43, 1257, 1998.
6. Moore, I.E., A key for the identification of animal hairs, *J. Forens. Sci. Soc.*, 28, 335, 1988.
7. Peabody, A.J. et al., The discrimination of cat and dog hairs, *J. Forens Sci. Soc.*, 23, 121, 1983.
8. Carracedo, A. et al., Isoelectric focusing patterns of some mammalian keratins, *J. Forens. Sci.*, 32, 93, 1987.
9. Jeffreys, A.J., Wilson, V., and Thein, S.L., Individual-specific 'fingerprints' of human DNA, *Nature*, 316, 76, 1985.
10. Jeffreys, A.J., Brookfield, J.F., and Semeonoff, R., Positive identification of an immigration test-case using human DNA fingerprints, *Nature*, 317, 818, 1985.
11. Jeffreys, A.J. and Morton, D.B., DNA fingerprints of dogs and cats, *Anim. Genet.*, 18, 1, 1987.
12. Gilbert, D.A. et al., Analytical DNA fingerprinting in lions: parentage, genetic diversity, and kinship, *J. Hered.*, 82, 378, 1991.
13. Menotti-Raymond, M. and O'Brien, S.J., Hypervariable genomic variation to reconstruct the natural history of populations: lessons from the big cats, *Electrophor.*, 16, 1771, 1995.
14. Tautz, D., Hypervariability of simple sequences as a general source for polymorphic DNA markers, *Nucl. Acids Res.*, 17, 6463, 1989.
15. Weber, J.L. and May, P.E., Abundant class of human DNA polymorphisms which can be typed using the polymerase chain reaction, *Am. J. Hum. Genet.*, 44, 388, 1989.
16. Sullivan, K.M. et al., Automated DNA profiling by fluorescent labeling of PCR products, *PCR Meth. Appl.*, 2, 34, 1992.
17. Ziegle, J.S. et al., Application of automated DNA sizing technology for genotyping microsatellite loci, *Genom.*, 14, 1026, 1992.
18. Micka, K.A. et al., Validation of multiplex polymorphic STR amplification sets developed for personal identification applications, *J. Forens. Sci.*, 41, 582, 1996.
19. Kimpton, C.P. et al., Validation of highly discriminating multiplex short tandem repeat amplification systems for individual identification, *Electrophor.*, 17, 1283, 1996.
20. Fregeau, C.J. and Fourney, R.M, DNA typing with fluorescently tagged short tandem repeats: a sensitive and accurate approach to human identification, *Biotechniq.*, 15, 100, 1993.
21. Kimpton, C.P. et al., Automated DNA profiling employing multiplex amplification of short tandem repeat loci, *PCR Meth. Appl.*, 3, 13, 1993.

22. Barendse, W. et al., A genetic linkage map of the bovine genome, *Nat. Genet.*, 6, 227, 1994.
23. Crawford, A.M. et al., An autosomal genetic linkage map of the sheep genome, *Genet.*, 140, 703, 1995.
24. Dietrich, W. et al., A genetic map of the mouse suitable for typing intraspecific crosses, *Genet.*, 131, 423, 1992.
25. Archibald, A.L. et al., The PIGMap consortium linkage map of the pig *(Sus scrofa), Mammal. Genome,* 6, 157, 1995.
26. Ellegren, H. et al., A primary linkage map of the porcine genome reveals a low rate of genetic recombination, *Genet.*, 137, 1089, 1994.
27. Jacob, H.J. et al., A genetic linkage map of laboratory rat, *Rattus norvegicus, Nature Genet.*, 9, 63, 1995.
28. Mellersh, C.S. et al., A linkage map of the canine genome, *Genom.*, 46, 326, 1997.
29. Menotti-Raymond, M. et al., A genetic linkage map of microsatellites in the domestic cat *(Felis catus), Genom.*, 57, 9, 1999.
30. O'Brien, S.J. et al., The Feline Genome Project, *Annu. Rev. Genet.*, 36, 657, 2002.
31. National Institutes of Health (active web site). http://www.Ncbi.nlm.hih.gov/entrez/query.fcgi?db=omia (On mendelian inheritance in animas; Retrieved May 3, 2007).
32. Menotti-Raymond, M. et al., Second generation integrated genetic linkage/radiation hybrid maps of the domestic cat, *J. Hered.*, 94, 95, 2003.
33. Menotti-Raymond, M. et al., Radiation hybrid mapping of 304 novel microsatellites in the domestic cat genome, *Cytogenet. Genome Res.*, 102, 272, 2003.
34. Murphy, W.J. et al., A radiation hybrid map of the cat genome: implications for comparative mapping, *Genome Res.*, 10, 691, 2000.
35. Murphy, W.J. et al., Development of a feline whole-genome radiation hybrid panel and comparative mapping of human chromosome 12 and 22 loci, *Genom.*, 57, 1, 1999.
36. Culver, M. et al., Genomic ancestry of the American puma (*Puma concolor*), *J. Hered.*, 91, 186–97, 2000.
37. Johnson, W.E. et al., Disparate phylogeographic patterns of molecular genetic variation in four closely related South American small cat species, *Mol. Ecol.*, 8, S79, 1999.
38. Eizirik, E. et al., Phylogeography, population history and conservation genetics of jaguars (*Panthera onca*), *Mol. Ecol.*, 10, 65, 2001.
39. Uphyrkina, O. et al., Conservation genetics of the Far Eastern leopard (*Panthera pardus orientalis*), *J. Hered.*, 93, 303, 2002.
40. Randi, E. et al., Genetic identification of wild and domestic cats (*Felis silvestris*) and their hybrids using Bayesian clustering methods, *Mol. Biol. Evol.*, 18, 1679, 2001.
41. Rueness, E.K. et al., Ecological and genetic spatial structuring in the Canadian lynx, *Nature,* 425, 69, 2003.

42. Driscoll, C.A. et al., Genomic microsatellites as evolutionary chronometers: a test in wild cats, *Genome Res.,* 12, 414, 2002.
43. Ernest, H.B. et al., Molecular tracking of mountain lions in the Yosemite valley region in California: genetic analysis using microsatellites and faecal DNA, *Mol. Ecol.,*9, 433, 2000.
44. Belden, R.C. and McCown, J.W., Florida Panther Reintroduction Feasability Study, Final Report: Florida game and Fresh Water Fish Commission, 1996.
45. Sores, T. et al., Paternity testing and behavioral ecology: A case study of jaguar (*Panthera onca*) in Emas National Park, Central Brazil, *Genet. Mol. Biol.*, 29, 735, 2006.
46. Higuchi, R. et al., DNA typing from single hairs, *Nature,* 332, 543, 1988.
47. Morin, P.A., Paternity exclusion using multiple hypervariable micorsatellite loci amplified from nuclear DNA of hair cells, in *Paternity in Primates: Genetic Tests and Theories,* Wickings, E.J., Ed., Basel: Karger, 1992, p. 63.
48. Menotti-Raymond M. et al., Genetic individualization of domestic cats using feline STR loci for forensic applications, *J. Forens. Sci.*, 42, 1039, 1997.
49. Menotti-Raymond, M.A., David, V.A., and O'Brien, S.J., Pet cat hair implicates murder suspect, *Nature,* 386, 774, 1997.
50. Hauge, X.Y. and Litt, M., A study of the origin of 'shadow bands' seen when typing dinucleotide repeat polymorphisms by the PCR, *Hum. Mol. Genet.*, 2, 411, 1993.
51. Interpretation of STR Profiles, Biology Section Methods Guide, 1996.
52. Jones, D.A., Blood samples: probability of discrimination, *J. Forens. Sci. Soc.*, 12, 355, 1972.
53. Budowle, B., Monson, K.L., and Chakraborty, R., Estimating minimum allele frequencies for DNA profile frequency estimates for PCR-based loci, *Int. J. Legal Med.*, 108, 173, 1996.
54. Menotti-Raymond, M. et al., An STR forensic typing system for genetic individualization of domestic cat (*Felis catus*) samples, *J. Foren. Sci.*, 50, 1061, 2005.
55. Butler, J.M. et al., The meowplex: a new DNA test using tetranucleotide STR markers for the domestic cat, *Prof. DNA*, 5, 7, 2002.
56. Coomber, N. et al., Validation of an STR multiplex typing system for genetic individualization of domestic cat (*Felis catus*) samples, (submitted 2007).
57. Fogle, B., *The New Encyclopedia of the Cat*, Rev. and updated ed., New York: DK Publishing, 2001.
58. Vella, C.M. et al., *Robinson's Genetics for Cat Breeders and Veterinarians,* 4th ed., Butterworth Heinemann, Edinburgh, 2003.
59. Menotti-Raymond, M., David, V.A., Weir, B., Coomber, N., and O'Brien, S.J., An STR population genetic database for genetic individualization of domestic cat (*Felis catus*) samples and initial validation studies, (in preparation).
60. DNA Advisory Board, Quality assurance standards for forensic DNA testing laboratories, *Forens. Sci. Comm.* 2 No. 3, July 2000.

61. Walsh, P.S., Erlich, H.A., and Higuchi, R., Preferential PCR amplification of alleles: mechanisms and solutions, *PCR Meth. Appl.*, 1, 241, 1992.
62. Wallin, J.M. et al., TWGDAM validation of the AmpFISTR blue PCR amplification kit for forensic casework analysis, *J. Forens. Sci.*, 43, 854, 1998.
63. Padar, Z. et al., Canine microsatellite polymorphisms as the resolution of an illegal animal death case in a Hungarian zoological gardens, *Int. J. Legal. Med.*, 115, 79, 2001.
64. Schneider, P.M., Seo, Y., and Rittner, C., Forensic mtDNA hair analysis excludes a dog from having caused a traffic accident, *Int. J. Legal Med.*, 112, 315, 1999.
65. Savolainen, P. et al., Sequence analysis of domestic dog mitochondrial DNA for forensic use, *J. Forens. Sci.*, 42, 593, 1997.
66. Menotti-Raymond, M., David, V., and O'Brien, S.J., DNA yield from single hairs (wool and guard/shed and plucked), success rate in amplifying STR and mtDNA targets, estimating DNA yield using multicopy target. Eleventh International Symposium on Human Identification. Biloxi, MS, 2000, http://www.promega.com/geneticidproc/ussvmpllproc/abstracts/menotti_ravmon d.pdf.
67. Budowle, B. et al., Mitochondrial DNA: a possible genetic material suitable for forensic analysis, in *Advances in Forensic Sciences*, Gaensslen, R.E., Ed., Medical Publishers, Chicago, 1990.
68. Menotti-Raymond, M. et al., Quantitative polymerase chain reaction-based assay for estimating DNA yield extracted from domestic cat specimens, *Croat. Med. J.*, 44, 327, 2003.
69. Morrison, T.B., Weis, J.J., and Wittwer, C.T., Quantification of low-copy transcripts by continuous SYBR Green I monitoring during amplification, *Biotechniq.*, 24, 954, 1998.
70. Yuhki, N. et al., Comparative genome organization of human, murine and feline MHC class II region, *Genome Res.*, 13, 1169–79, 2003.
71. Lander, E.S. et al., Initial sequencing and analysis of the human genome, *Nature*, 409, 860, 2001.
72. Waterston, R.H. et al., Initial sequencing and comparative analysis of the mouse genome, *Nature*, 420, 520, 2002.
73. Walker, J.A., Hughs, D., and Batzer, M., SINE based PCR for the identification of species-specific DNA, Thirteenth International Symposium on Human Identification, Phoenix, Arizona, 2002.
74. Mandrekar, M.N. et al., Development of a human DNA quantitation system, *Croat. Med. J.*, 42, 336, 2001.
75. Holland, M.M. and Parsons, T.J., Mitochondrial DNA sequence analysis—validation and use for forensic casework, *Forens. Sci. Rev.*, 11, 1, 1999.
76. Gill, P. et al., Identification of the remains of the Romanov family by DNA analysis, *Nat. Genet.*, 6, 130, 1994.

77. Holland, M.M. et al., Mitochondrial DNA sequence analysis of human skeletal remains: identification of remains from the Vietnam War, *Forens. Sc.L.*, 38, 542, 1993.
78. Paabo, S., Gifford, J.A., and Wilson, A.C., Mitochondrial DNA sequences from a 7000-year old brain, *Nucl. Acids Res.*, 16, 9775, 1988.
79. Cash, H.D., Hoyle, J.W., and Sutton, A.J., Development under extreme conditions: forensic bioinformatics in the wake of the World Trade Center disaster, *Pac. Symp. Biocomput.*, 638, 2003.
80. Lopez, J.V., Cevario, S., and O'Brien, S.J., Complete nucleotide sequence of the domestic cat *(Felis catus)* mitochondrial genome and a transposed mtDNA tandem repeat *(Numf)* in the nuclear genome, *Genom.*, 33, 229, 1996.
81. Lopez, J.V. et al., Numt, a recent transfer and tandem amplification of mitochondrial DNA to the nuclear genome of the domestic cat, *J. Mol. Evol.*, 39, 174, 1994.
82. Fridez, F., Analyse D'ADN mitochondrial animal: vers une exploitation forensique des poils d'animaux domestiques, 1'Institut de Police Scientifique et de Criminologie, Lausanne: 1'Universite de Lausanne, 1999, p. 199.
83. Eizirik, E. et al., Molecular genetics and evolution of melanism in the cat family, *Curr. Biol.*, 13, 448, 2003.
84. Schmidt-Kunzel, A. et al., Tyrosinase and tyrosinase related protein 1 genetic variants specify domestic cat coat color alleles of the albino and brown loci, *J. Hered.*, 96, 289, 2005.
85. Ishida, Y. et al., A homozygous single-base deletion in MLPH causes the dilute coat color phenotype in the domestic cat, *Genomics*, 88, 698, 2006.
86. Wright, S., The albino series of allelomorphs in guinea-pigs, *Am. Nat.*, 49, 140, 1915.
87. Imes D.L. et al., Albinism in the domestic cat (*Felis catus*) is associated with a tyrosinase (TYR) mutation, *Anim. Genet.*, 37, 175, 2006.
88. Lyons L.A. et al., Tyrosinase mutations associated with Siamese and Burmese patterns in the domestic cat (*Felis catus*), *Anim. Genet.*, 36, 119, 2005.
89. Balding, D.J., When can a DNA profile be regarded as unique? *Sci. Justice*, 39, 257, 1999.
90. Boxall, B., Pet DNA gaining value as evidence in crimes, *Los Angeles Times*, 2001.
91. The Associated Press, Prosecutors use dog's DNA to convict man of stabbing London bouncer, StarTribune.com, 2001.
92. The Associated Press, Dog DNA convicts man, *Genet.*, 1, 1, 2001.

An Overview of Insect Evidence in Forensic Science

6

HEATHER MILLER COYLE, PH.D.

Contents

Forensic entomology is a subspecialty of forensic science that uses insect evidence for legal applications. The role of insect evidence at crime scenes typically is a supporting one, as this form of evidence is rarely used without additional case information and evidence. There are three major arenas in which forensic entomology is best applied: (1) estimating postmortem interval in cases of human death, (2) detection of narcotics from insects recovered from decomposing human remains, and (3) contamination of food by insect infestation (i.e., civil casework). This brief article provides an overview of each of the three applications.

Forensic entomologists cannot pinpoint time of death to an accurate minute or hour. Rather, insect information from a human corpse can provide a minimum estimation of the postmortem interval (PMI) (Figures 1 and 2). Insect species have been studied extensively in the laboratory and entomologists know under "perfect" growth conditions how long it would take for an insect species to hatch and mature into each subsequent stage of growth. By adding these time intervals together, one can determine the time required for an insect to develop from egg to adult. These estimates are minimum time interval estimates, since many factors can influence the time of growth at each stage of the insect's life cycle. These factors include ambient temperature and weather conditions such as rain and humidity; whether the crime scene is indoors or outdoors; and whether the body is covered or uncovered. The forensic investigator should take note of ants, butterflies, flies, larvae, maggots, bees, wasps, and any insect body parts (Figure 3), as their presence may be of importance to the forensic entomologist. These types of insects feed on proteins and sugar-rich body fluids subsequent to death.

STAGES OF DECAY:
FRESH STAGE – moment of death to bloated appearance
BLOATED STAGE – abdomen inflated, fluid seepage from orifaces, soil analysis for fluids and microbes
DECAY STAGE – skin breakage, gases escape, corpse deflates
POST-DECAY STAGE – approximately 10% of corpse tissue mass remains
SKELETAL STAGE – bones, teeth, no insects on body

Figure 1 Stages of human and animal decay.

Insect maggots may be of different species and are often difficult to identify until they mature into adults. When collecting maggots, each should be measured in length and preserved in separate vials containing ethyl alcohol. The region and condition of the body at time of collection should also be well documented. Some of the live maggots can be taken to the laboratory to rear into adulthood for identification purposes. Sometimes, the maggot species and time intervals give conflicting estimates of postmortem interval since death. When this situation occurs, it is important to consider microclimates that may affect time interval interpretation. A good example of when this situation might occur is when a body is found partially submerged in water. As the body tissue dries, it becomes more difficult for maggots to feed; therefore, one time estimate may be accurate for the drying portion of the

Insects Collected from a Decaying Corpse
FRESH STAGE – blow flies and flesh flies
BLOATED STAGE – house flies, beetles; e.g., rove, hister, burying
DECAY STAGE – arthropods and mites, large masses of feeding maggots
POST-DECAY – primarily beetles
SKELETAL STAGE – insects, bacteria, etc. found in soil below and can remain for several years as signs of decomposition

Figure 2 Insect species that may be collected from a decomposing body.

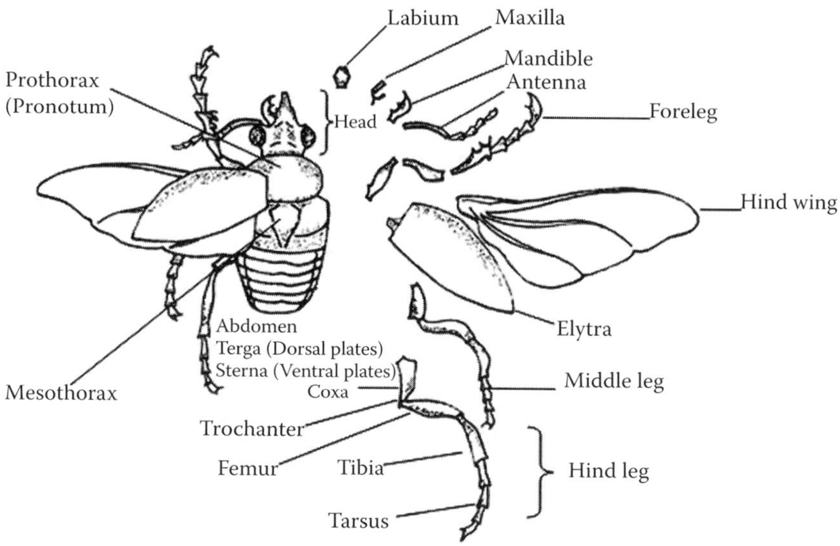

Figure 3 Diagram of generalized insect external anatomy. (Illustration courtesy of John E. Coyle.)

body, while the soft submerged tissue may allow for a longer feeding time for that population of maggots on the body. A forensic entomologist will also consider, when determining the time estimate, how long a species might take on average to colonize a body after death; sometimes this adds several days to the postmortem interval estimate. DNA testing of maggot crop contents has been used to identify missing persons as well as for interpreting PMI. A comparison of preservatives has been conducted and 70–95% ethanol was found to be the best for later DNA recovery from the maggot. Kahle's solution and formaldehyde were detrimental to DNA recovery and fresh-frozen maggots stored at –70°C were optimal.

Entomological evidence recovered from decomposing human remains has been studied for the ability to detect narcotics transferred from the body to the insect via insect feeding activity. Some initial studies seem promising for the collection and analysis of feeding insects for determination of presence or absence of narcotics in the deceased person at time of death. This, of course, would be a supplement or replacement to the existing toxicological screening tests routinely used in forensic science and would be particularly useful if the body is not located immediately after death (i.e., highly decomposed). It was thought that insects could "concentrate" narcotics to a detectable level even when they could not be identified by standard tests from urine, blood, or tissue. Some substances that have been identified from insects feeding on corpses include: phenobarbital, cocaine, malathion, mercury, heroin, methamphetamine, amitryptyline, and phencyclidine. It is also important to know if a person may have ingested narcotics prior to death, since some drugs and their metabolites can also alter insect growth stages, an added factor in using insects to calculate postmortem intervals. Many synthetic compounds have not been studied adequately, along with their effects on human decomposition and insect feeding and growth, and so additional research is needed.

Not all insect evidence is related to crime scenes; however, much entomological evidence is associated with civil matters. A certain number of insect parts or remnants are tolerated by the Food and Drug Administration (FDA) and United States Department of Agriculture (USDA) as part of the processing and packaging of premade food and food products for human consumption. The allowable level of insect matter tolerated in food is determined by a published table of values called Defect Action Level (DAL); if the insect matter exceeds the DAL, then the food must be recalled for further evaluation. Some forensic entomologists serve as consultants for major food and beverage corporations to assist in the identification and eradication of such insect pests in food items. Barcodes and lot numbers on food items allow one to track the original harvest, process plant, manufacturer, and storage facilities to help pinpoint the time frame for food contamination. In addition, forensic entomologists may be involved indirectly in real estate and home owner sales disputes where cockroach and termite infestations occur. Entomologists may be asked to aid in establishing the length of time an insect colony has been present within a building structure and assessing whether a home owner could reasonably have had prior knowledge of the insects and building damage before the real estate sale. Bedbugs have recently increased in numbers and infestations for hotels and motels throughout the world. Although they do not transmit any disease to humans, they do feed on sleeping human hosts and leave visible bites and blood trails on their victims. Forensic entomologists may be called on to evaluate infestations and cases of neglect in hospitals and nursing homes.

DNA testing has been useful for the identification of many insect species in forensic science; although the test may be somewhat costly, it does provide a rapid means for identification with especially limited samples. Since insect cases and fragments may be useful evidence, the PCR amplification step can expand the ability of the forensic entomologist to make an insect or even human identification. The application of human body lice as a forensic tool has been investigated and proven to be useful to link the DNA profile of the assailant to the victim via the transfer of lice during close contact assaults. The DNA profiles were detectable up to twenty hours postfeeding. Case studies with maggots have been performed with STR or HVR analysis from homicides to associate adult flies with a specific corpse (human STR tests) as well as fly species identification by analysis of the cytochrome b gene.

The time of storage and PMI in these cases appeared to not influence the quality of the DNA results.

Bibliography

Tomberlin, J.K., Wallace, J.R., and Byrd, J.H., Forensic entomology: myths busted! *Forens. Mag.*, 3, 10–14, 2006.

Goff, M.L., *A Fly for the Prosecution: How Insect Evidence Helps Solve Crimes*, Harvard University Press, Cambridge, MA, 2000.

Amendt, J., Krettek, R., and Zehner, R., Forensic entomology, *Naturwissenschaften*, 91, 51–65, 2004.

Mumcuoglu, K.Y., Gallili, N., Reshef, A., Brauner, P., and Grant, H., Use of human lice in forensic entomolog, *J. Med. Entomol.*, 41, 803–806, 2004.

Linville, J.G., Hayes, J., and Wells, J.D., Mitochondrial DNA and STR analyses of maggot crop contents: effect of specimen preservation technique, *J. Forens. Sci.*, 49, 341–344, 2004.

Zehner, R., Amendt, J., and Krettek, R., STR typing of human DNA from fly larvae fed on decomposing bodies, *J. Forens. Sci.*, 49, 337–340, 2004.

Campobasso, C.P., Linville, J.G., Wells, J.D., and Introna, F., Forensic genetic analysis of insect gut contents, *Am. J. Forens. Med. Pathol.*, 26, 161–165, 2005.

Harvey, M.L., An alternative for the extraction and storage of DNA from insects in forensic entomology, *J. Forens. Sci.*, 50, 627–629, 2005.

Chen, W.Y., Hung, T.H., and Shiao, S.F., Molecular identification of forensically important blow fly species (Diptera: Calliphoridae) in Taiwan, *J. Med. Entomol.*, 41, 47–57, 2004.

7 Use of Forensic DNA Typing in Wildlife Investigations

RICHARD M. JOBIN, PH.D.

Contents

7.1 Introduction

DNA analysis (in general) and analysis of microsatellite or short tandem repeat (STR) fragments of DNA (in particular) have been used in a wide variety of wild and domestic animals. STR-based DNA typing tests have been developed for just about every conceivable species of wildlife ranging from the insects to whales.[1,2] STR analysis of wildlife species has been used to study population structure (genetic differences between populations of animals) and also to ascertain evolutionary relationships between groups of animals of the same species or groups from different species.[3,4]

In the field of forensics, in either humans or wildlife, the development of DNA typing represents the largest single advance in technology since the discovery of the forensic value of fingerprints. The power of discrimination between individuals when using reasonably polymorphic STR loci combined with the high sensitivity of the polymerase chain reaction (PCR) enables investigators to link miniscule, sometimes microscopic, amounts of biological material back to victims or perpetrators. STR analysis is in routine use for humans in a forensic capacity and several commercial kits are available for this purpose. Most major federal and local law enforcement agencies use these commercial human identification STR-based kits. Several STR markers have been developed and/or tested for large game and fish species and are routinely used in the investigation of wildlife offenses.[5–7] Our own lab has developed DNA typing tests for mule deer, white-tailed deer, moose, bighorn sheep, cougar, grizzly bear, and black bear.[8,9] Some of these STR markers may be available in commercial kits (now or soon), but most laboratories performing wildlife testing are constructing their own tests and kit components.

7.2 Why Use This Technology for Wildlife Forensics?

Illegal poaching of animals is a serious problem; in cases in which there is a black market for animal parts, some species are driven to the point of extirpation. Black rhinos, elephants, and walleye (a North American game fish) are all species whose populations have been seriously impacted by poaching.[10,11] However, even in species that are more abundant such as North American deer, significant numbers of animals are taken illegally.[12,13] This level of poaching becomes an issue when opportunities for observing or harvesting these animals is reduced. Rather than punish legal hunters by reducing quotas, implementing draws, or closing certain areas to hunting, it is more reasonable to concentrate efforts on the illegal taking of these resources.

Unfortunately, wildlife offenses, such as poaching, are very difficult to detect and prosecute. This problem stems from the fact that there are relatively few wildlife officers available to cover large areas of undeveloped land. Because these offenses occur in remote areas, it is also unlikely that there will be any witnesses to any given offense. These circumstances make it quite difficult to detect wildlife offenses and make it even more difficult to convict individuals who commit these offenses. Contrast this situation with that of a police officer investigating a homicide or assault. In human investigations, there is usually less terrain to cover, so an officer has (1) less distance to travel to get to the crime scene and (2) is more likely to encounter criminal activity while on patrol. Due to higher population levels in developed areas, there is also a much greater likelihood that there will be witnesses to any given event. Additionally, if the incident under investigation is not fatal, the officer may be able to get valuable information by interviewing the victim.

Forensic DNA typing is an especially useful tool for wildlife investigations and the case can be made that this technology is, in fact, even better suited for wildlife investigations than it is for investigations into human cases of murder or assault. This can be explained by the following rationale: in homicide or assault cases, the motivation for these crimes is generally financial or emotional and the perpetrator usually has no reason to purposefully collect biological material from the victim. In crimes against humans, biological material may be accidentally transferred between the perpetrator and victim during the commission of a violent act. This becomes valuable evidence due to Locard's principle of exchange. In wildlife cases the scenario is quite different; here the actual purpose for committing the offense is to collect biological material from the victim (e.g., a deer). The collection of this material may be in the form of meat for consumption, a trophy for display, or various other body parts that may have medicinal or cultural significance. In addition to the biological material that a perpetrator purposefully collects, other biological material will unintentionally become associated with the perpetrator. When butchering or preparing a big game animal in a remote location (often in the dark) it is almost impossible for the perpetrator to avoid incidental transfer of blood, hair, and tissue to their clothing, belongings, and vehicle. In addition to the intentional and accidental accumulation of biological material, perpetrators or wildlife offenses usually leave the less valuable parts of large game (such as bones and viscera) at the kill site. In cases in which the game animal is particularly large, it may be impossible to move the entire animal from the field unless it is cut into smaller pieces. This situation results in a considerable amount of biological material being left at the kill site. Evidence, such as bones, may remain at the kill site for more than a year. DNA typing can then be used to link meat found in a freezer to remains left at a kill site across a considerable span of time.

There is no question that forensic DNA typing is an effective technology to use in wildlife investigations; however, is the effectiveness of this technique worth the increased cost of analysis? It is the experience of our own laboratory and many other police laboratories that the use of forensic DNA results in a higher proportion of guilty pleas.[14] These pleas save not only officer time, but also court time. In fact, it stands to reason that the earlier in an investigation the DNA results are made available, the less costly the investigation and court process. This creates a convincing argument for the adequate staffing and funding of crime labs so that prompt results can always be made available. Additionally, if a crime laboratory is visible in a community, officers may use the lab as "verbal leverage" when interviewing suspects and may lead to a prompt confession from the suspect. This represents the greatest cost savings because the investigative effort is minimized and no analysis is required from the laboratory. A more difficult effect to measure is the deterrent effect that a DNA typing program has on wildlife offenses. If a laboratory's capabilities and successes are well publicized, it may well serve to deter a certain percentage of would-be perpetrators of wildlife offenses. Habitual and/or opportunistic poachers may reconsider their behavior if they know that it can be detected even in the absence of an eyewitness.

7.3 Operational Considerations

Implementation of forensic DNA typing into an enforcement operation has a profound effect on all aspects of the program, including laboratory function, case management, and evidence collection. PCR-based DNA typing is a very sensitive technique that also has a very high level of discrimination. The strength of this new technique, coupled with the fact that in many jurisdictions wildlife penalties have increased in severity,[15] has raised the stakes so that fish and wildlife officers and their associated laboratories are now being held to similar standards as for those agencies that investigate and prosecute human crime. In some cases, animal DNA evidence has been rejected by courts because the submitting laboratory did not treat the data in a way that was similar enough to the way that human DNA is treated.[16] Additionally, there is also the possibility that animal DNA (e.g., especially cat and dog) will be used as evidence in a human trial.[17] The above circumstances require that personnel in the field and laboratory take special care and follow appropriate protocols when handling DNA evidence.

7.3.1 Evidence Collection

The high level of sensitivity of PCR-based DNA typing necessitates the altering of evidence collection protocols so that the likelihood of cross-contamination is reduced. In the field, officers must use proper leak-proof packaging, fresh instruments, and fresh gloves when collecting evidence. Additionally, officers should collect evidence that has a lower DNA content, such as trace evidence, before handling evidence that has a higher DNA content, such as a carcass. For example, in the case of a bear mauling, an officer should first collect evidence from the victim, where he is searching for trace evidence (bear hair), and then collect a sample from the trapped or dead bear. This greatly reduces the possibility of introducing hair from the suspect bear to the victim. Similar evidence handling advice has been offered for collection of DNA samples in human crimes.[18] Wildlife officers would be well advised to follow this guideline when collecting DNA evidence during wildlife investigations.

7.3.2 Laboratory Organization

As is the case for evidence collection, due to the sensitivity of PCR-based DNA typing, special procedures also must be followed in the laboratory. Forensic DNA labs should be designed so that separate areas are used for: 1) evidence examination (searching), 2) DNA extraction, 3) general reagent and PCR preparation, and 4) DNA amplification and sizing. PCR enclosures (e.g., laminar flow hoods) with ultraviolet (UV) sterilization are useful equipment for keeping work surfaces free of ambient or background DNA. Using the enclosures for the preparation of PCR reactions (i.e., master mix consisting of PCR primers, buffer, and enzyme) and the addition of DNA to the master mix will help to reduce the probability of contaminating samples. Additionally, questioned and known samples should always be handled separately. The traditional wisdom, from labs that handle human samples, has always been that questioned samples should also always be handled before known samples (i.e., sample processing for evidence is separated in time and/or space from known reference sample processing). This action reduces the chance that a known sample that usually has a high DNA content will contaminate a questioned sample, which may have a low DNA content. However, these definitions and practices can significantly break down in wildlife investigations due to the nature of the cases.

7.3.3 Questioned Versus Known Samples in Wildlife Investigations

In wildlife investigations, determining which samples are of questioned origin and which samples are of known origin can be more complicated than one might expect. In cases involving humans, this categorization is quite simple. Known samples are obtained from the suspect and/or victim and are of known origin. These samples are often obtained by an officer via a DNA warrant. On the other hand, when the investigator does not know the origin of the sample, it is categorized as a questioned sample. These samples are often collected at the scene of the crime. In wildlife cases, evidence rarely falls into these neat categories. For example, in a case where a sample is collected from a "gut pile" left in a provincial park and another sample is a bloodstain on a suspect's vehicle, which sample is questioned and which is known? Even if your suspect has an entire carcass in his possession and a sample is collected from it, do you have a known sample from a victim? If a human body is found hanging in a garage, there is no doubt that an illegal act has occurred. However, a big game carcass found hanging in a garage may be perfectly legal. Therefore, one may argue that in wildlife investigations, all samples are questioned samples. It may be more appropriate to label samples in wildlife investigations as "incident" samples and "suspect" samples. Incident samples can be defined as biological material from wildlife that is found in association with an area where a violation of wildlife statutesis alleged to have occurred or a place where a violent episode occurred. Examples of this type of sample include blood in a field, viscera at a kill site, and bear hair in the clothing of a victim. Suspect samples can be defined as biological material from wildlife that is found in association with an individual suspected of committing wildlife offenses. Examples of this type of sample include a carcass at a suspect's residence, blood on a suspect's vehicle, and a sample from a bear suspected of mauling a human.

The question remains, when analyzing DNA for wildlife investigations, which evidence is processed first, scene or suspect samples? The analyst should always handle and analyze the samples that have the lowest DNA content first because they are much less likely to contaminate a sample that has a higher DNA content. If DNA from one sample contaminates another, the amount of DNA being transferred from one sample to the next is usually quite low. Therefore, if the sample that has received the contamination already has a high level of its own (native) DNA, then it is likely that the signal from the contaminating DNA will be "drowned out" or overwhelmed by the native DNA (the DNA that is supposed to be there and what you are trying to detect). However, if the DNA content in the sample receiving the contamination has a low native DNA content, then it is much more likely that the signal from the contaminating DNA will be detected. Therefore, the analyst would be well advised to process the evidence that has the lowest DNA content first, regardless of the origin of the sample. If both sample types have comparable DNA levels, there is no scientific rationale for a preference in the order of the processing of the samples. However, in our laboratory, incident samples are always processed first unless the suspect samples clearly have a lower DNA content. This procedure is followed so that all cases received by the lab are handled in a consistent manner. As a rule, evidence can be separated into three different categories based on their DNA content: 1) high DNA content (an ample amount of muscle tissue, organs, or antler), 2) moderate DNA content (an ample amount of blood, hair, or bone), 3) trace DNA content (very limited amounts of blood, hair, saliva, or bone). These categories can be used to determine the order of evidence processing.

In summary, evidence collected in the course of wildlife investigations can be categorized as being associated with the incident or suspect. To avoid problems with contamination, incident and suspect samples should always be handled and processed separately. Furthermore, the DNA content of the evidence should determine whether incident samples are handled before suspect samples or vice versa. The type of sample (incident or suspect) that has the lowest predicted DNA content based on the type of sample should be handled and processed first.

7.4 Analytical Considerations

Most DNA labs have a very similar flow for the processing of exhibits for analysis. DNA analysis usually includes the following steps:

Exhibit searching/identification of biological substances
Identification of species
Extraction of DNA
Quantification of DNA
Amplification of DNA
Sizing of DNA fragments
Interpretation of DNA profiles
Statistical treatment of genetic data
Production of a report/providing court testimony

7.4.1 Exhibit Searching/Identification of Biological Substances

Generally, two tasks are performed at this stage of case processing. The first task is to locate biological material and take an appropriate sample of this material for further analysis. The second task is to make an initial assessment of the nature of the biological material. Location of biological material can be very simple when dealing with exhibits where such material is abundant. Examples of such exhibits would include an animal carcass or a piece of meat. Conversely, finding trace evidence, such as bear hair or saliva on the clothing or belongings of a mauling victim, can be quite challenging. Regardless of the nature of the biological material or exhibit being examined, it is of the utmost importance that the forensic biologist conducts the search in a methodical and thorough manner. It is equally important that proper records are made during the examination. Notes made during the examination should include case number, exhibit number, date, signature, description of the exhibit, description of the exhibit packaging, records of samples taken from the exhibit, and results of preliminary tests and assessments (see example of exhibit worksheet, Figure 1).

EXHIBIT WORKSHEET

ALBERTA FISH AND WILDLIFE FORENSIC LABORATORY

Case Number:
Exhibit Number:

Contributor:

Name:	District:	Province:

Caption:

Species:	Offence:	Suspect/Accused:

Description of packaging:

Description of Exhibit(s):

Notes:

Date:
Signature:

Figure 1 Example of exhibit worksheet.

An initial assessment of the biological material usually includes identification of the biological material(s) present. These may include blood, saliva, feces, hair, hide, bone, antler, feather, scales, and various internal tissues and organs. These identifications are based on macroscopic and microscopic characteristics and in the case of blood are determined using a commercially prepared product, Hemastix® (Bayer Inc. Healthcare Division, Toronto, Ontario), for a presumptive test[19] and/or the hemochromogen test as a confirmatory test.[20] Furthermore, species identification may be possible if the morphology of the biological material is unique to a particular species. For example, the morphology of a moose antler is unique to that particular species.

7.4.2 Identification of Species

In contrast to investigations involving human crime, wildlife investigations may include several game species. Since different species of animals often require the use of different DNA typing tests, it is necessary to identify the species from which the biological material originates before initiating STR typing. The species of origin of biological material can be determined through analysis of proteins[21] or via DNA analysis[22] (usually using DNA sequence data).

7.4.3 Extraction of DNA

There are numerous methods for extracting DNA from different biological materials. These extraction methods include phenol chloroform extraction,[23] guanidine isothiocyanate extraction,[24] manual commercial kits such as the Qiagen Dneasy® tissue kits (Qiagen Inc., Valencia, CA, USA), and commercial kits for magnetic bead extractors such as the Promega MagneSil KF, Genomic System (Promega Corporation, Madison, WI, USA). Depending on the condition of your exhibit and what type of biological material is present, any of the above extraction techniques may be the most effective. In wildlife investigations, forensic biologists will examine a wide variety of biological materials that originate from many species and that have suffered a wide variety of environmental insults. It is not uncommon to analyze evidence that has been cooked, smoked, tanned, or left out of doors for prolonged periods of time. Each lab should experiment with different extraction techniques to determine which techniques and conditions work most effectively for their particular situation.

7.4.4 Quantification of DNA

Once extracted, DNA is usually quantified so that the forensic examiner can add the appropriate amount to the PCR reaction. In human forensics, techniques such as DNA probes or real-time PCR are used to specifically quantify the amount of human DNA present in the sample.[25,26] In North America there are over 15 big game species that can be legally hunted; this does not include fish species that can be legally caught and mammals that can be legally trapped. The large number of species that many wildlife laboratories handle makes it impractical to develop special tests to quantify DNA for each species. Fortunately, use of a simple UV spectrophotometer has proven to be an effective method to quantify DNA. Although this instrument cannot distinguish between DNA from different species, the experience of our laboratory is that a vast majority of PCR reactions receive an appropriate amount of DNA for proper amplification.

7.4.5 Amplification of DNA

Most wildlife laboratories amplify DNA from short tandem repeat (STR) loci. Although these loci are less polymorphic than variable number of tandem repeat (VNTR) loci, they are more numerous, more robust, and more amenable to automation and are therefore excellent candidates as forensic markers.[27] Use of different-colored fluorescent tags on primers allows the forensic examiner to simultaneously interrogate several loci, even if their alleles have overlapping size ranges (see Figure 2). The majority of alleles that are utilized in forensic analysis are larger than 60 base pairs and smaller than 500 base pairs. The basic components of a PCR reaction include: (1) PCR Buffer (commercially available), (2) magnesium chloride ($MgCl_2$), (3) deoxyribonucleotide triphosphates (dNTPs), (4) at least one set of PCR primers, (5) Taq polymerase, and (6) template DNA. Depending on the examiner's needs, the amounts of components are varied or additional components may be added.[28] Likewise, cycling parameters chosen by

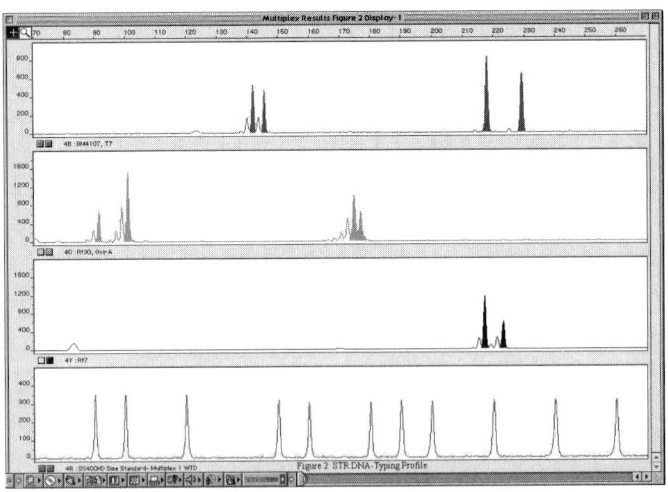

Figure 2 STR profile from a wildlife case sample. (**See color insert after p. 108.**)

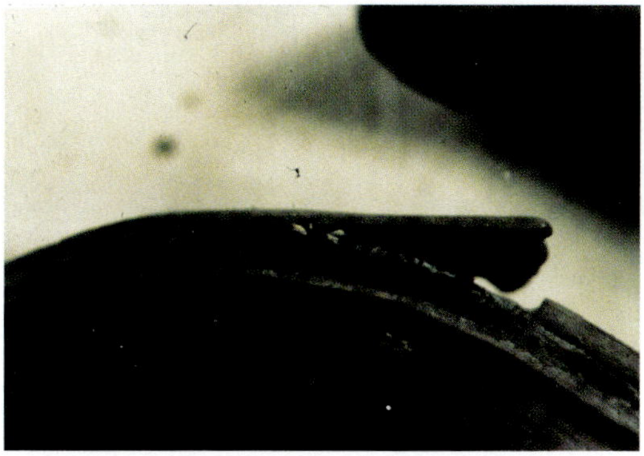

Color Figure 2.1 Some grass seeds (located in the center of the photo, bottom of boot) that were recovered from the suspect's shoes ultimately matched back to those collected from the crime scene and proved to be valuable associative evidence.

Color Figure 2.2 Plant material that was collected from the crime scene is shown in this photo. Seed pods are visible and the seeds from the suspect's boots were consistent with coming from the same source plant.

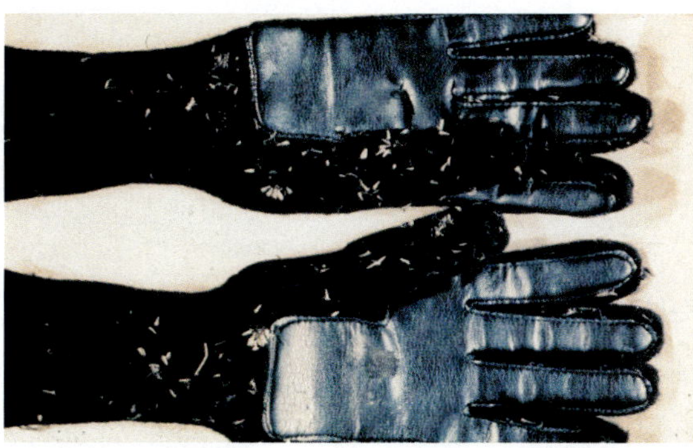

Color Figure 2.3 The suspect's gloves collected from the crime scene and sent to the forensic science laboratory were covered in numerous grass seeds. A simple microscopic comparison can add extra information to show a linkage between individuals and the crime scene.

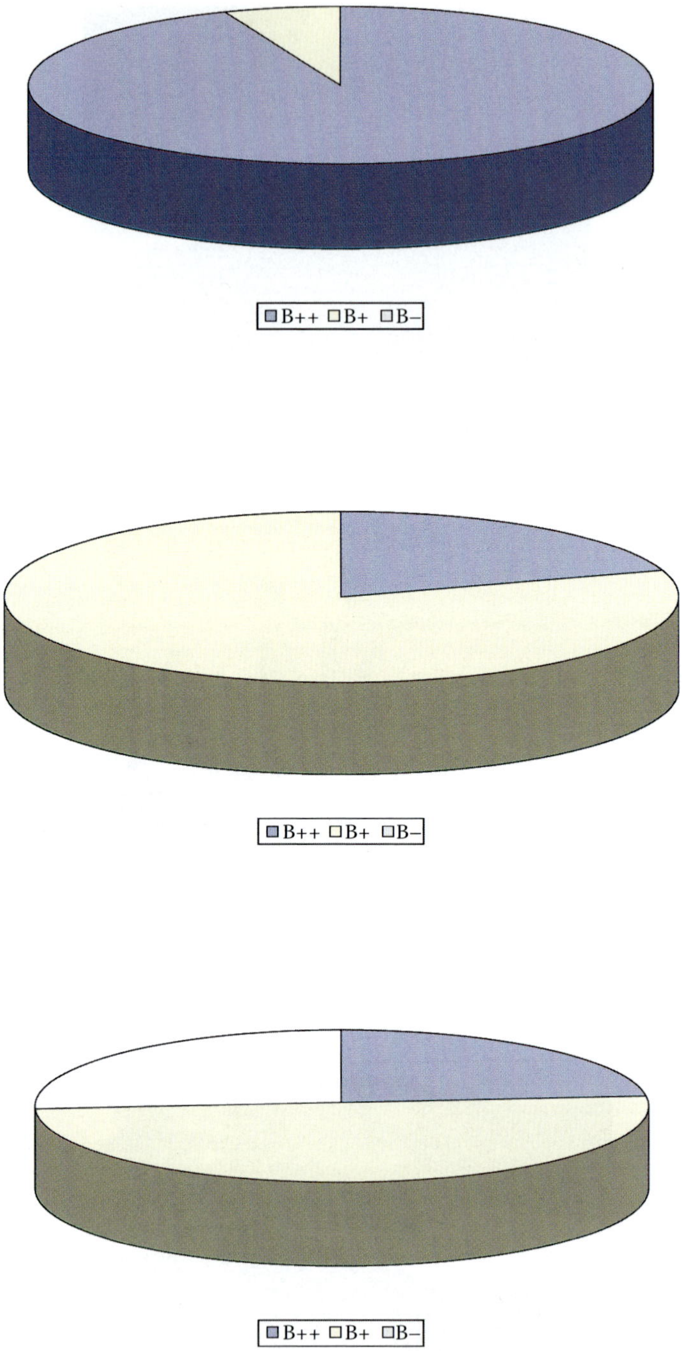

Color Figure 4.5 Success rate of canine-specific STR profiling depending on the amount of visible blood on the sampling device. The three categories "full profiles," "partial profiles," and "no results" are distinguished. The 52 samples were grouped into three classes: (B++), much blood (n = 23); (B+), little blood (n = 21); (B–), no blood (n = 8).

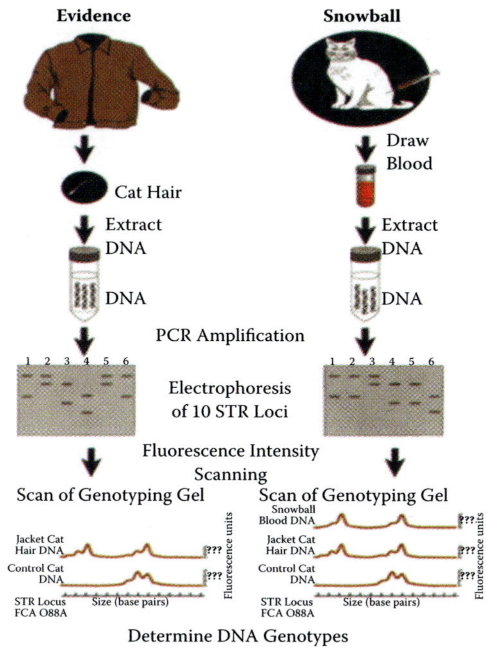

Color Figure 5.4 Graphics presented to the Prince Edward Island jury depicting the sequence of steps in generating DNA profiles for evidentiary hairs and blood sample from Snowball.

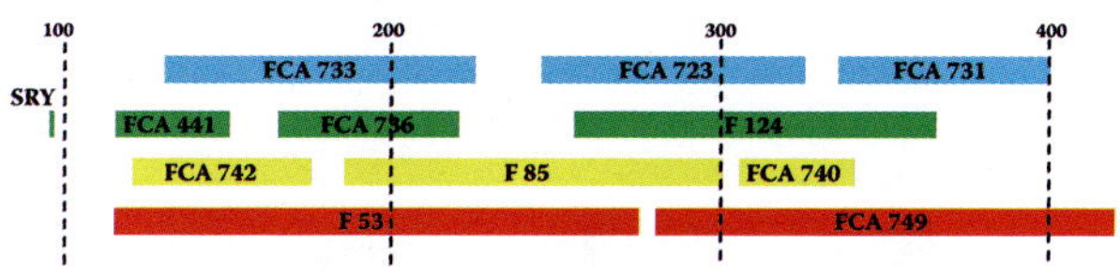

Color Figure 5.6 Fluorescent dyes and size ranges for domestic cat 12-plex as observed in a 1,043- member domestic cat genetic database generated from 38 cat breeds (Menotti-Raymond et.al., in preparation).

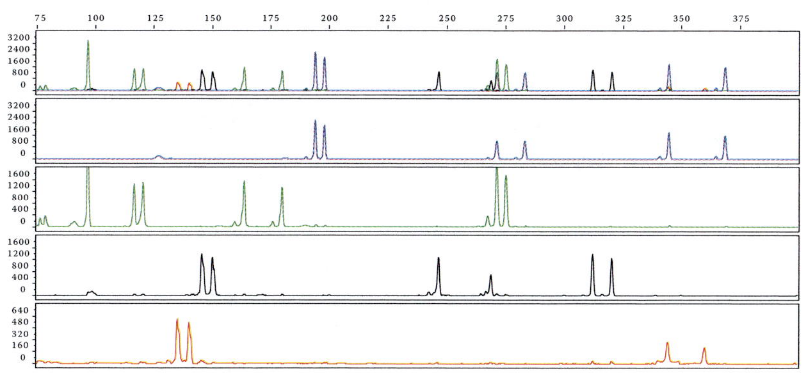

Color Figure 5.7 Electrophoretogram of PCR products of 12-member multiplex amplified from 4 ng of male genomic DNA (upper panel). Lower panels demonstrate PCR products labeled with fluorescent tags FAM (3 STR), VIC (3 STR, SRY gene), NED (3 STR), and PET (2 STR).

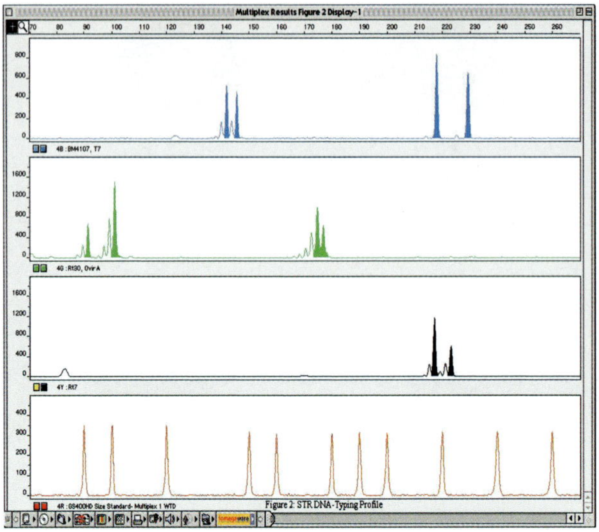

Color Figure 7.2 STR profile from a wildlife case sample.

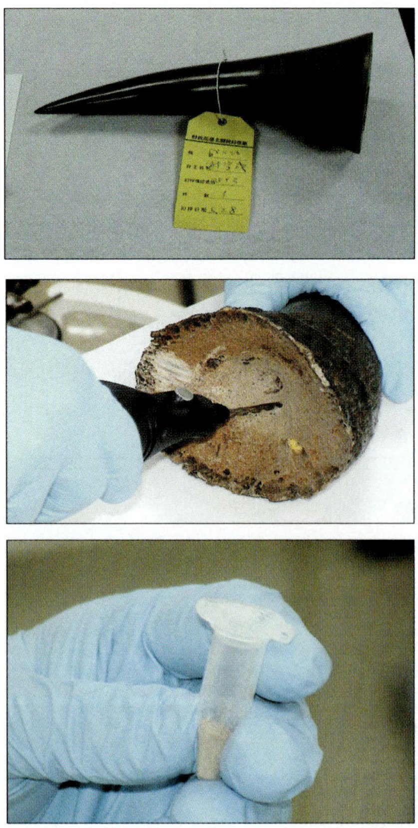

Color Figure 8.1 DNA sampling from animal horn. We use a low-speed electric drill to make powder from the sample.

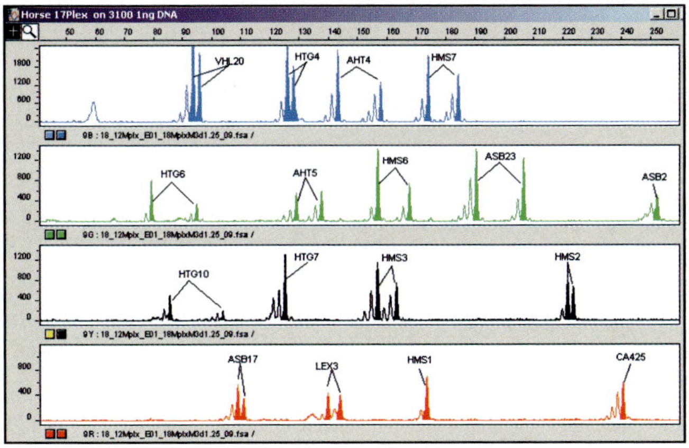

Color Figure 8.2 GeneScan electropherograms of amplified control horse DNA by 17-Plex Horse Genotyping kit run on ABI Prizm 3100 DNA Analyzer, [DNA] = 1.25 ng.

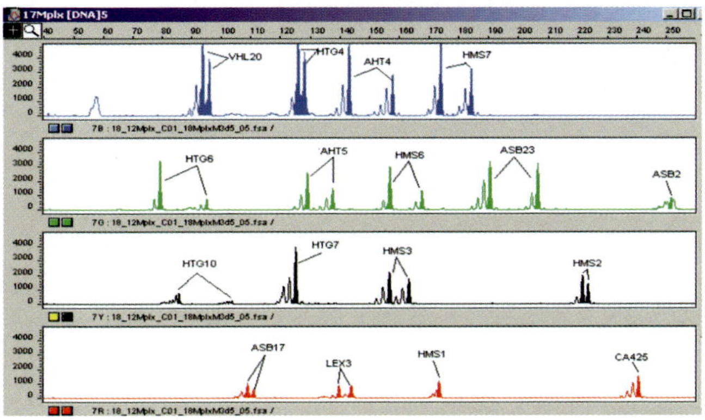

Color Figure 8.3 GeneScan electropherogram of amplified control horse DNA by 17-Plex Horse Genotyping kit run on ABI3100 instrument, [DNA] = 5 ng.

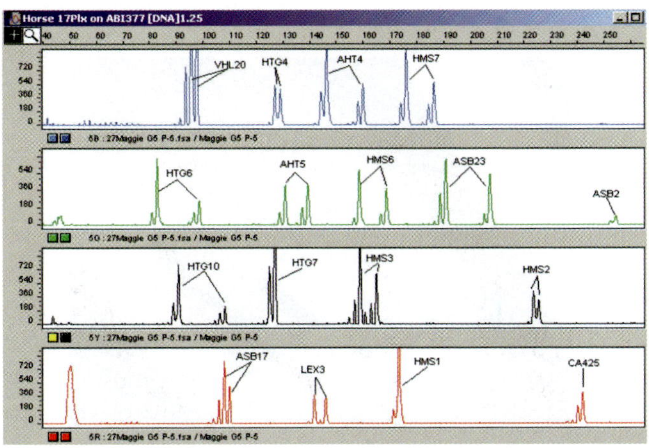

Color Figure 8.4 GeneScan electropherogram of amplified control horse. DNA by 17-Plex Horses Genotyping kit run on ABI377 instrument, [DNA] = 1.25 ng.

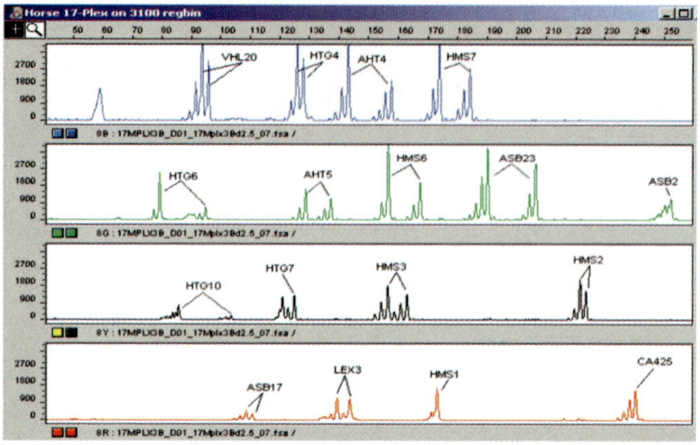

Color Figure 8.5 GeneScan electropherogram of amplified horse DNA by 17-Plex Horses Genotyping kit run on 3100, [DNA] = 2.5 ng, by using GeneScan36_POP4DefaultModule.

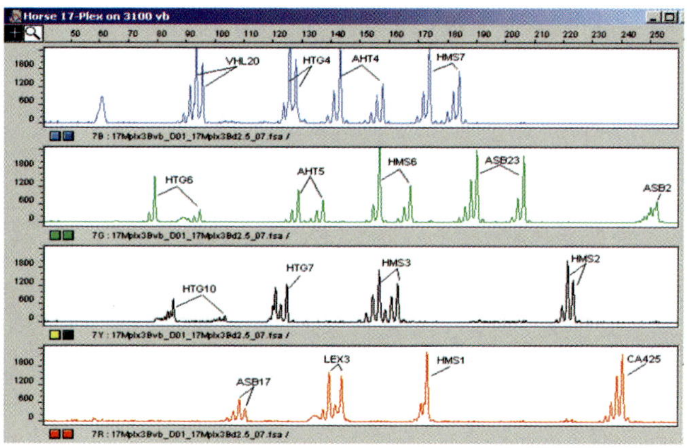

Color Figure 8.6 GeneScan electropherogram of amplified horse DNA by 17-Plex Horses Genotyping kit genotyped on 3100 run, [DNA] = 2.5 ng, by using GeneScan36vb_-POP4DefaultModule.

Color Figure 8.7 Exotic pets like chameleons are difficult to care for but highly desired by pet stores and exotic animal lovers alike.(Photograph is courtesy of John E. Coyle.)

Color Figure 8.8 Tortoises are sold as pets but their shells and meat are also sold as delicacies in many Asian countries, leading to a steady decline in their population sizes. (Photograph is courtesy of John E. Coyle.)

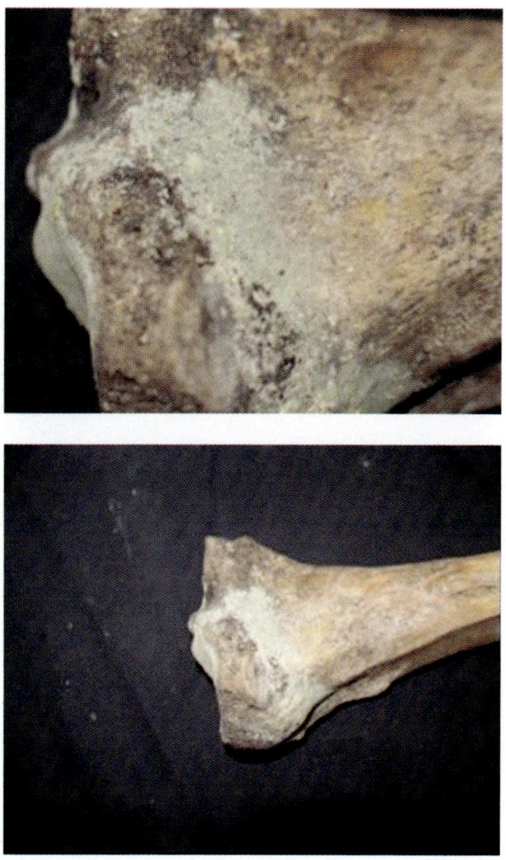

Color Figure 9.2 Fungus growth on human tibia, proximal end, posteriorly. (Courtesy of International Commission on Missing Persons, Bosnia & Herzegovina)

(a) (b)

Figure 9.3 (a) Fungus growth on chicken bone (scale: 1 mm) kept in a closed container with some moisture at room temperature for few weeks, (b) a close-up photo of some fungal fruiting bodies.

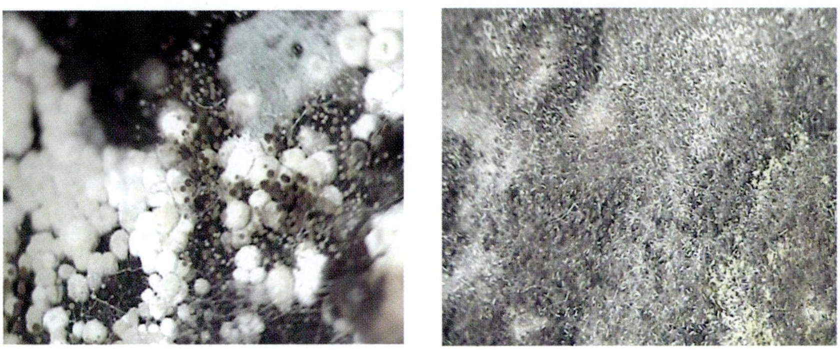

(a) (b)

Figure 9.4 Fungi growth on animal liver tissue: (a) after several days, and (b) after a few months. The sample was kept inside a closed plastic container at room temperature.

a particular laboratory may vary; however, most protocols include the following steps: (1) denaturation, (2) annealing, and (3) extension.[29] Development of multiplexed reactions can save analyst time and reduce the amount of consumables used during analysis. A typical PCR reaction that would be used in our laboratory includes GeneAmp® PCR Buffer II (Roche Molecular Systems Inc., Almeda, CA, USA) @ 1X, $MgCl_2$ @1.3–2.5 mM; GeneAmp® dNTP mix with dTTP @ 0.2 mM (Roche Molecular Systems Inc., Almeda, CA, USA); one to seven sets of fluorescently tagged primers @ 0.08–1.00 μm; AmpliTaq Gold® (Roche Molecular Systems Inc., Almeda, CA, USA) @ 1–3 Units; Template DNA @ 5–30 ng; and filtered, autoclaved, and deionized water added to make a final 25-μL reaction volume. A typical cycling program used in our laboratory includes: (1) hot start cycle at 95°C for 4 minutes, (2) denaturing cycle at 94°C for 30 seconds, (3) annealing cycle at 40–60°C for 30 seconds, (4) extension cycle at 72°C for 1 minute, (5) steps 2–4 are repeated 30–40 times, (6) final extension at 60°C for 45 minutes, (7) hold at 22°C overnight, and (8) store amplicons at –20°C. There are many manufacturers and models of thermal cyclers to choose from. Since there are few commercial kits available for nonhuman DNA analysis, wildlife forensic analysis often requires a substantial amount of development. It is therefore advisable to purchase a thermal cycler that has "gradient" capability. This feature allows the researcher to run several annealing temperatures and magnesium concentrations simultaneously, reducing the time spent on optimizing primers.

STRs occur in various sizes of repeats ranging from two to seven repeats that generate PCR products that are generally less than 400 base pairs in length.[30] Commercial kits that are used for human forensics primarily use loci with tetranucleotide repeats. This selection of loci with larger-size repeats reduces the amount of stutter observed relative to the size of the true peak. The reduced stutter size facilitates the interpretation of mixed profiles.[31] Dinucleotide repeats have more prominent stutter. Although this complicates interpretation of mixtures, the prominent stutter serves as a diagnostic marker, allowing the forensic examiner to distinguish true peaks from artifacts. This feature is quite useful when working with noncommercially produced tests that tend to have more artifactual noise.

7.4.6 Sizing of DNA Fragments

The two most popular technologies for the sizing of DNA fragments are "slab gel" electrophoresis and capillary electrophoresis. New DNA chip technology is still under development.[32] Slab gels, capillaries, and chips use the same basic principles. Amplified DNA fragments are separated by size and charge through the use of a gelatinous matrix and electrical charge. Along with the

DNA fragments, a size standard is also run and used to estimate the size of the amplified product. DNA fragments are sized via software such as ABI Prism and GeneScan® analysis software (Applied Biosystems, Foster City, CA, USA). Peaks are assessed as being true alleles based on their magnitude, morphology, and size.

7.4.7 Interpretation of DNA Profiles

DNA profiles are interpreted in relation to database information that has been previously collected. Samples from populations of approximately 100 individuals are collected and analyzed from locations across the geographic range of the species being studied. These populations of animals are genotyped at a number of selected loci. At each locus, alleles of different sizes are grouped or binned according to their size. This categorization is accomplished via software such as ABI Prism and Genotyper® analysis software (Applied Biosystems, Foster City, CA, USA). Due to the repetitive nature of STRs, the categories tend to be evenly distributed in a punctuated fashion (i.e., each category is separated by two base pairs in dinucleotide repeats). Numbers of alleles occurring in each size category are counted to determine the allele frequencies at each locus. In most instances, at each locus, some alleles will be more common (have a high frequency of occurrence) while others will be more rare (having a low frequency of occurrence). In casework, incident and suspect samples are analyzed and the amplified DNA fragments are assigned to specific size categories. A match can be declared only if all of the amplified DNA fragments from all of the loci that are being tested in the incident and suspect samples are of the same size (fall in the same categories). Standard samples are run along with the casework samples to ensure that the DNA fragments are sizing correctly and being assigned to the appropriate categories. The standard sample is a sample that is part of the database for that species. The analyst can then determine if the database sample that is run during the case is being categorized in the same way it was when it was run as part of the database. If there is a discrepancy between the two samples the categories can then be offset or the case can be reanalyzed.

7.4.8 Statistical Treatment of the Genetic Data

As in most other areas of forensic DNA analysis, human DNA has led the way in setting precedents and standards for the statistical treatment of forensic DNA typing data. In fact, the issue of statistical treatment of forensic DNA evidence in humans has arguably been subjected to more scrutiny than any other subject in the field of forensic science. Many committees and conferences have examined this subject and two landmark publications have been the result of these efforts.[33,34] Most enforcement laboratories follow the recommendations set out in these publications. Since the final objective of forensic analysis is court acceptance, it would be highly advisable to adopt the statistical methodologies that have been previously accepted in your particular court system and to change methodologies as prescribed by the courts. Failure to do so may result in the court rejecting your work.[16] Our laboratory has used the same statistical software (Genescan®, Genotyper®, and Genepop version 3.2a) (Genepop is available free of charge at "ftp://isem. isem.univ-montp2.fr/pub/pc/ genepop/" and many of the same NRC II prescribed measures that are used by the Royal Canadian Mounted Police. These measures include database populations of 100–120 individuals, conservative estimates of minimum allele frequencies, and adjustments for population subdivision. At points where our laboratory differs from court-accepted treatment of the data, we invariably use even more conservative treatments. For example, rather than using the recommended frequency estimate of homozygous genotypes (the square of the frequency of the homozygous allele, $P = p^2$[34] we use the unmodified frequency of the homozygous allele ($P = p$). We adopted this measure because we encounter null alleles with a greater frequency than that experienced in human DNA typing. Also, following the lead of local law enforcement, our laboratory provides estimates of the rarity of the genetic profile in question as random match probabilities. As a result of the above, our laboratory can inform the courts that, in preparation of our casework, we use the same statistical treatment of the data that the court has already repeatedly accepted. Furthermore, any departures from this court-accepted treatment of the data results in a more conservative estimate of the random match probability, which works in the defendant's favor.

7.4.9 Production of a Report/Providing Court Testimony

The information that has been gained through the forensic examination and analysis of the exhibits is then presented to the court in the form of a forensic laboratory report. This report should contain the following information:

Lab case number
Case number from the client
Suspect name
Name of the submitting officer
Date that the case was received at the laboratory

A list of the exhibits received
Descriptions of the biological material identified
Descriptions of the comparisons of the DNA typing profiles produced by the exhibits examined and the resultant matches and exclusions between the exhibits
The current disposition of the exhibits
The forensic examiner's name, signature, and contact information.

Figure 3 is a representative forensic laboratory report.

FISH AND WILDLIFE FORENSIC LABORATORY
O.S. LONGMAN BLG. 6909-116 STREET EDMONTON ALBERTA CANADA T6H 4P2

Alberta
SUSTAINABLE RESOURCE DEVELOPMENT

FORENSIC LABORATORY REPORT

This report may not be published, in whole or in part, without the written consent of the author. Page 1 of 2

LAB CASE NUMBER: 04-000 **DISTRICT NUMBER: 04-0000**

DATE RECEIVED: 2004-April-01 **DATE REPORTED:2004-April-14**

SUSPECT/ACCUSED: I.M. POACHER

To: Ike Ketchem
Your Town, Alberta

Exhibits Examined:

Scene Exhibits

Exhibit 295-101[Red Substance-Scene]
Exhibit 295-102[Hair-Scene]

Suspect Exhibit

Exhibit 295-201[Meat WTD Female-Carcass]

Forensic Analysis:

Examinations were conducted to determine any forensically significant associations among biological materials obtained from exhibits.

Analysis of deoxyribonucleic acid using polymerase chain reaction amplification of several loci followed by measurement by electrophoresis was performed.

Conclusions:

1. Hair found in the questioned exhibit, **Exhibit 295-102[Hair-Scene]** was identified as being from the **Deer Family**.

2. The red substance found in questioned exhibit **Exhibit 295-101[Red Substance-Scene]**was positively identified as **blood**.

Figure 3 Example of a forensic wildlife report.

FISH AND WILDLIFE FORENSIC LABORATORY
O.S. LONGMAN BLG. 6909-116 STREET EDMONTON ALBERTA CANADA T6H 4P2

Alberta
SUSTAINABLE RESOURCE DEVELOPMENT

FORENSIC LABORATORY REPORT

This report may not be published, in whole or in part, without the written consent of the author. Page 2 of 2

3. **Exhibit 295-102[Hair-Scene]** and **Exhibit 295-101[Red Substance-Scene]** produced the same DNA-typing profile. This profile, produced from questioned exhibits, also **matches** the DNA-typing profile obtained from the known exhibit, **Exhibit 295-201[Meat WTD Female-Carcass]**. The estimated probability of selecting an unrelated individual at random from the Your Town White-tailed Deer population with a matching profile is **1 in 10 Billion.**

Remarks:

The following loci were amplified and analyzed using Applied Biosystems reagents and capillary electrophoresis instrumentation and an MJ Research thermalcycler: Multiplex 1 (BM4107, T7, OvirA, Rt30, Rt7); Multiplex 2 (Rt5, BM1225, OheN, BM4208, OheQ).

All exhibits are being retained at the Alberta Fish and Wildlife Forensic Laboratory and will be disposed of one year after the report date unless otherwise advised.

Prepared by:

Richard M. Jobin Ph.D.
Forensic DNA Specialist

Figure 3 (Continued)

7.5 Summary

Because wildlife offenses occur in remote locations, they are difficult to detect and prosecute. However, these offenses usually result in the perpetrator collecting and possessing large amounts of biological material. This material may be in the form of intentionally collected meat or trophies such as antlers and can also include accidentally accumulated blood and hair. The perpetrator also often leaves a large amount of biological material at the kill site. The above circumstances make forensic DNA typing a highly effective tool for resolving wildlife cases. Additionally, if forensic DNA results are available to officers at a relatively early stage in the investigation, the use of this technology reduces the overall cost of investigating and prosecuting these cases. Gaining court acceptance of forensic DNA typing results can be greatly simplified if forensic wildlife biologists follow the same protocols, analytical techniques, and statistical treatment of the data that has been used in the forensic DNA analysis of humans. In addition to forensic DNA typing being an extremely valuable enforcement tool, data collected by gathering and analyzing forensic databases may also provide valuable biological information that may be useful in the management of game animals.

7.6 An Example of the Use of Forensic DNA Typing in Casework

Case details: A landowner phoned in a complaint indicating that a truck was on his property chasing deer and that he had chased the truck off with his own vehicle but the truck was waiting on a nearby road. Officers responded, detained the suspects, and began their investigation. The officers discovered a white-tailed deer doe in the truck's box and upon examining the deer, found that it had received considerable trauma. They also discovered blood, hair, and two deer fetuses in the landowner's field and that the field had been "churned up" by a vehicle. The suspect's vehicle was seized and transported to a secure government warehouse. Forensic staff were called in to perform a vehicle search and to examine the deer carcass. The carcass had evidence of blunt trauma but no evidence of gunshot wounds. There was a large hole in the lower abdomen of the deer and no uterus was present. There was no indication that the deer had recently delivered young vaginally. The doe was missing a great deal of hair from one side but there was no indication of disease or poor health. Upon searching the vehicle, hair was found on the front grill, sides, box, and under the vehicle. Blood was also found under the vehicle and in its box. The suspect maintained that the deer was road kill that he picked up to use for bear bait.The evidence was submitted to the forensic laboratory for further examination and analysis. The hair found in the field and on the truck was identified as being from the deer family. DNA was extracted from the evidence and amplified in two multiplexed PCR reactions, which provided a 10-locus DNA profile. The genotypes produced from the blood and hair found in the field, as well as the profiles produced from the blood and hair found on the vehicle, matched the White-tailed deer carcass at all loci. The random match probability for this particular genotype was found to be one in 10 quadrillion. Furthermore, the two fetuses from the field were also genotyped and found to have DNA profiles that were consistent with being the offspring of the deceased female deer.

References

1. Keyghobadi, K., Roland, J., and Strobeck, C., Isolation of novel microsatellite loci in the Rocky Mountain apollo butterfly, *Parnassius smintheus, Heriditas.*, 136, 247, 2002.
2. Berube, M. et al., Polymorphic di-nucleotide microsatellie loci isolated from the humpback whale, *Megaptera novaeangliae, Mol. Ecol.*, 9, 2181, 2000.
3. Angers, B. et al., Specific microsatellite loci for brook charr reveal strong population subdivision on a microgeographic scale, *J. Fish Biol.*, 47 (Suppl. A), 177, 1995.
4. Petren, K., Grant, B.R., and Grant, P.R., A phylogeny of Darwin's finches based on microsatellite DNA length variation, *Proc. R. Soc. Lond. B*, 266, 321, 1999.

5. Ellegren, H., Anderson, L., and Wallin, K., DNA polymorphism in the moose (*Alces alces*) revealed by the polynucleotide probe (TC)n, *J. Hered.*, 82, 429, 1991.
6. DeWoody, J.A., Honeycutt, R.L., and Skow, L.C., Microsatellite markers in white-tailed deer, *J. Hered.*, 86, 317, 1995.
7. Jones, K.C., Levine, K.F., and Banks, J.D., DNA-based markers in black-tailed and mule deer for forensic applications, *Calif. Fish Game*, 86, 115, 2000.
8. Packer, T., Patterson, D., and Jobin, R.M., Forensic DNA typing in grizzly bears in the Province of Alberta. *Can. Soc. Forens. Sci. J.*, 36, 65, 2003.
9. Jobin, R.M., Patterson, D., and Stang, C., Forensic DNA typing in several big game animals in the Province of Alberta, *Can. Soc. Forens. Sci. J.*, 36, 56, 2003.
10. Messer, K., The poacher's dilemma: the economics of poaching and enforcement, *Endang. Spec. Update,* 17, 50, 2000.
11. Sullivan, M.G., Illegal angling harvest of walleyes protected by length limits in Alberta, *North Am. J. Fish. Manag.*, 22, 1053, 2002.
12. McCorquodale, S.M., Movements, survival and mortality of black-tailed deer in Klickitat Basin of Washington, *J. Wildl. Manage.*, 63, 861, 1999.
13. Fuller, T.K., Dynamics of a declining white-tailed deer population in north-central Minnesota, *Wildlife Monogr.*, 110, 1, 1990.
14. Parks, S.A., Compelled DNA testing in rape cases: illustrating the necessity of an exception to the self-incrimination clause, *Wm. & Mary J. Women & L.*, 7, 499, 2001.
15. Servetnyk, R., Operation Tamerack, *Alberta Game Ward.*, Fall, 1, 2002.
16. "http://www.courts.wa.gov/opinions/?fa=opinions.opindisp&docid=435078 MAJ" (accessed Aug. 2004).
17. Menotti-Raymond, M., David, V.A., and O'Brien, S.J., Pet cat hair implicates murder suspect, *Nature*, 386, 774, 1997.
18. Wickenheiser, R.A., Trace DNA: a review, discussion of theory, and application of the transfer of trace quantities of DNA through skin contact, *J. Forens. Sci.*, 47, 442, 2002.
19. Product information sheet for Hemastix®. Product number 2524A, Bayer Inc. Healthcare Division. Toronto, Ontario M9W 1G6.
20. Kerr, D.J.A. and Mason, V.H., The haemochromogen crystal test for blood, *Brit. Med. J.*, 1, 134, 1926.
21. Packer, T.C.D. et al., Identification of bear and cervid species with double immuno-diffusion and isoelectric focusing, 12th International Symposium on Human Identification, Biloxi, MS, Oct 8–9, 2001.
22. Paetkau, D. and Weldon, J., Identification of mammalian species through partial (A-only) DNA sequencing of the 16S rRNA gene, *Conserv. Genet.*, submitted.
23. Gross-Bellard, M., Oudet, P., and Chambon, P., Isolation of high-molecular-weight DNA from mammalian cells, *Eur. J. Biochem.*, 36, 32, 1973.
24. Boom, R. et al., Rapid and simple method of purification of nucleic acids, *J. Clin. Microbiol.*, 28, 495, 1990.

25. Andreasson, H., Gyllensten, U., and Allen, M., Real-time DNA quantification of nuclear and mitochondrial DNA in forensic analysis, *Biotechniq.*, 33, 407, 2002.
26. Waye, J.S. et al., Sensitive and specific quantification of human genomic deoxyribonucleic acid (DNA) in forensic science specimens: casework examples, *J. Forens. Sci.*, 36, 1198, 1991.
27. Hammond, H.A. et al., Evaluation of 13 short tandem repeat loci for use in personal identification applications, *Am. J. Hum. Genet.*, 55, 175, 1994.
28. Harris, S. and Jones, D.B., Optimization of the polymerase chain reaction, *Br. J. Biomed. Sci.*, 54, 166, 1997.
29. Saiki, R.K., The design and optimization of PCR, in *PCR Technology: Principles and Applications for DNA Amplification*, Erlich, H.A., Ed., W.H. Freeman and Company, New York, 1992, pp. 7–16.
30. Wallin, J.M. et al., TWGDAM validation of Ampf*l*str blue PCR amplification kit for forensic casework, *J. Forens. Sci.*, 43, 854, 1998.
31. Walsh, P.S., Fildes, N.J., and Reynolds, R., Sequence analysis and characterization of stutter products at the tetranucleotide repeat locus vWA, *Nucl. Acids Res.*, 24, 2807, 1996.
32. Munro, N.J. et al., Molecular diagnostics on microfabricated electronic devices: from slab gel- to capillary- to microchip-based assays for T- and B-cell lymphoproliferative disorders, *Clin. Chem.*, 45, 1906, 1999.
33. Grossblatt, N., Ed., *DNA Technology in Forensic Science*, National Academy Press, Washington, D.C., 1992.
34. Grossblatt, N., Ed., *The Evaluation of Forensic DNA Evidence*, National Academy Press, Washington, D.C., 1996.

DNA Testing of Animal Evidence — Case Examples and Method Development

8

EN-UEI JIANG, W.A. VAN HAERINGEN, PH.D.,
L.H.P. VAN DE GOOR, PERO DIMSOSKI, PH.D.,
AND HEATHER MILLER COYLE, PH.D.

Contents

8.1 Animal DNA Identification Service in MJIB, Taiwan, R.O.C.

En-Uei Jiang

Taiwan often encounters problems when it comes to identifying wildlife and their products. For the sake of protecting endangered wildlife, the DNA Laboratory under the authority of the Ministry of Justice Investigation Bureau (MJIB) is assigned to offer animal DNA identification services. A traditional way of identification is determining the species by assessing animals' appearances according to morphology. When it comes to animal products (e.g., meat or horns) that can't be judged by appearances, methods such as protein electrophoresis, immunology testing, and high-performance liquid chromatography (HPLC) analysis are used. However, the existence of some proteins is not the same for all cells. Instead, it varies in accordance with cell differentiation. Degradation of proteins also can happen shortly after animals die; therefore, identification based on proteins often fails.

Recently, the use of molecular biology in the identification of species has successfully resolved the problems described above. For each creature, every cell carries the same DNA (deoxyribonucleic acid) code. Therefore, the examination of blood, muscle tissues, bones, fur, or even processed products would produce the same DNA result. Furthermore, the identification of DNA requires significantly less quantity of sample than that needed for protein analysis. Reproducing certain fragments of trace DNA permits the examination of a species and individualization of a sample.

MJIB has been developing identification techniques based on DNA Cyt*b* (cytochrome *b*) sequencing analysis. The nucleotide sequence of Cytb in DNA is moderately conserved across animal species; however, sufficient genetic variation exists, that is, several bases of difference can be found even among the more evolutionarily closely related species of animals. Therefore, the result can be a basis for animal species identification. After two years of research and validation, this technology has been used on actual cases since 1998. An average of 40–50 cases are taken each year, which involves mostly imported or smuggled animal products from Southeast Asia.

From years of experience with animal identification in our country, 90% of wildlife products are fake or falsely advertised. For example, cow and pig bones are often advertised as representing the following meats or powders: tiger bones, the tendon of a cow, the penis of a tiger, a buffalo horn, or a rhinoceros horn. A cobra gall can actually be a more common snake gall, chicken gall, or duck gall. In a 2003 case, examination of meat of so-called rare animals from a specialist store revealed that the meat claimed to be from highly valued white-nosed badgers was actually that of lower valued pigs and ostriches. In June 2004, a customs agent found an animal horn brought into Taiwan by a traveler. He claimed it to be a yak horn; however, after the customs sent this horn to be identified, it was found to be a rhinoceros horn (Figure 1). The horn is drilled and processed for DNA much like human bone samples.

In addition, female birds have ZW (heterozygous) sex chromosomes while male birds have ZZ (homozygous) ones. Therefore, the gender of birds is decided by the W chromosome from female birds, which is the opposite of mammals. Most birds also do not have obvious external sex organs like mammals. Although some grown birds have different feather colors, appearances, and behaviors that make it easier to judge their gender, females and males of many bird species share the same appearance. Therefore, differentiating the

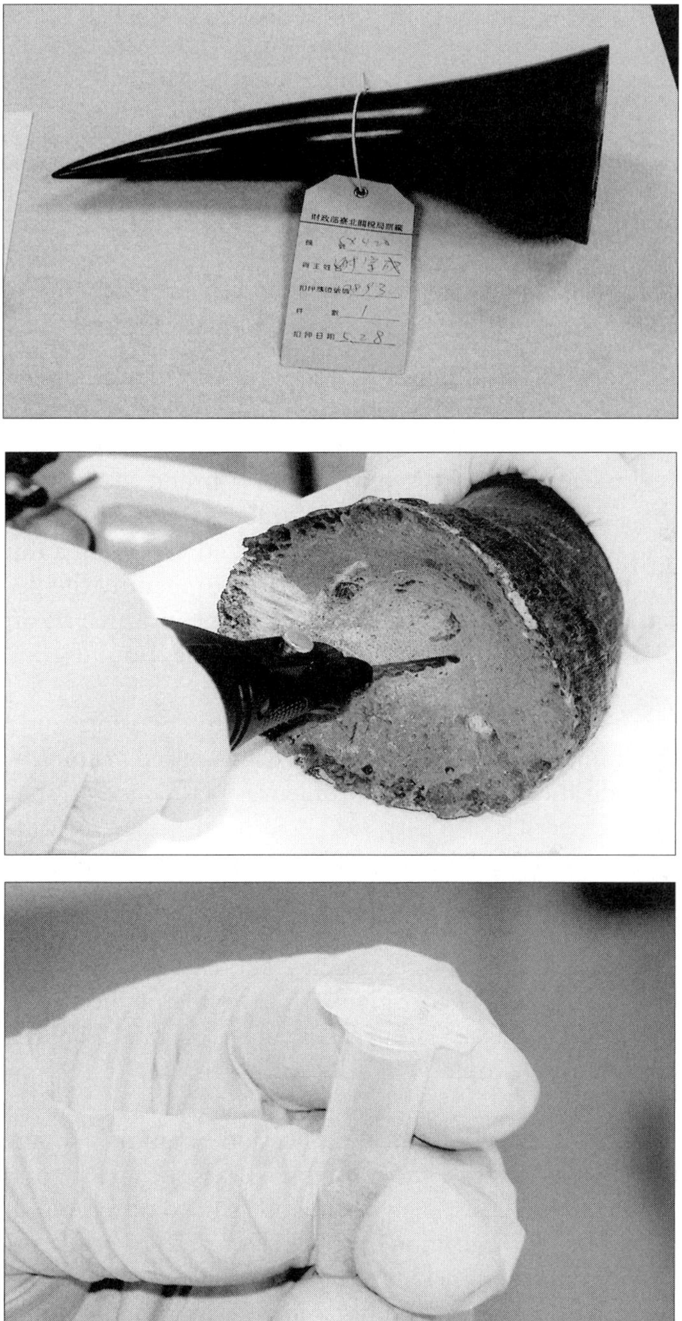

Figure 1 DNA sampling from animal horn. (**See color insert after p. 108.**) We use a low-speed electric drill to make powder from the sample.

gender of birds is very difficult morphologically. Furthermore, determining the gender of chicks makes it easier to keep track of and conserve them in the future in breeding programs. Consequently, MJIB has also developed DNA techniques for the identification of the gender of birds.

8.2 Identification of Horses Sold for Slaughter Using 17 Genetic Markers

W.A. van Haeringen Ph.D. and L.H.P. van de Goor

Horses with a top-level pedigree are sometimes sold for slaughter for various reasons, mostly being that such horses have poor racing health. Over the past years, five horses that were "sold for slaughter" turned up again after a period of time. New owners tell of the presence of a brand identifying horses as registered to a specific organization or studbook. Consequently, when the new owner inquires about the pedigree associated with such a horse bought from a horse trader, DNA analysis can be used to identify the pedigree of a horse. This is, of course, only possible in situations in which an organization requires a DNA profile from each registered horse, resulting in an equine DNA database.

DNA was extracted from equine hair root samples using our laboratory's standard operating procedures. Multiplex PCR was used to amplify 17 genetic markers.[1] Visualization of the DNA profile was performed on the ABI Prism 3100 DNA Sequencer (Foster City, CA) Microsoft Access software was used to search the equine DNA sample profile for matching DNA profiles in our reference horse databases.

In all cases submitted, we could identify the original DNA profile, notwithstanding the fact that before March 2003, DNA analysis was based on only 12 genetic markers. The original DNA profile of the newly identified horses could easily be found. Confirmation of the original stored sample with the newly submitted sample was done using the set of 17 microsatellite markers.

DNA analysis focusing on the identification of animals should use minimum standards.[2] In sensitive and emotional cases such as the one described here, the quality of the DNA analysis must be sufficient to hold up in court. Over the past two years, a transition from the classical set of 9 to 12 microsatellite markers[3] has been made to include a set of 17 microsatellite markers[1] that are routinely typed in the laboratory for horse identification and individualization. The use of microsatellite markers is extremely powerful for the identification of horses. As previously described,[3,4] a minimum set of 10 variable microsatellite markers should be sufficient for testing and individualizing an equine sample.

References

1. Dimsoski, P., Development of a 17-plex microsatellite polymerase chain reaction kit for genotyping horses, *Croat. Med. J.*, 44, 332–335, 2003.
2. Budowle, B. et al., Recommendations for animal DNA forensic and identity testing, *Int. J. Legal. Med.*, 16, epub ahead of print.
3. Bowling, A.T. et al., Validation of microsatellite markers for routine horse parentage testing,. *Anim. Genet.*, 28, 247–252, 1997.
4. Tozaki, T. et al., Population study and validation of paternity testing for thoroughbred horses by 15 microsatellite loci, *J. Vet. Met. Sci.*, 63, 1191–1197, 2001.

8.3 Lineage Analysis and Identification of Pigeons

W.A. van Haeringen Ph.D. and L.H.P. van de Goor

In relation to a criminal case, pigeons were intended as a means to transport ransom. To do so, a blackmailer handed over several pigeons, to which diamonds were to be attached before releasing them to fly back to the pigeon's owner. However, using a radio device, the pigeons were followed to the suspected criminal. A large number of feather samples were then used to identify relationships as well as identity among the limited number of pigeons used for the exchange of ransom, in connection with feathers found at the site where the pigeons flew (suspect's residence). The evidence generated by the DNA analyses of feathers and pigeons proved to be of major importance in this case.

DNA was extracted from feathers using our laboratory's standard operating procedures. Multiplex PCR was used to amplify 10 genetic microsatellite markers. Visualization of the resultant DNA profile was performed on the ABI Prism 3100 DNA Sequencer (Foster City, CA).

Microsoft Access software was used to search the database for matching DNA profiles.

The 10 genetic markers revealed a large number of alleles in the pigeons typed (Table 1).

In total, four pigeons were submitted for the transfer of ransom. These four pigeons were compared to approximately 150 feather samples found in the car of the suspect and the area where the four pigeons landed. The comparison was aimed at establishing pigeon identity and lineage. Feathers found in the car matched one of the four pigeons; lineage analysis identified several direct parental relationships among the four pigeons and the feathers found in the area where the pigeons landed. Forensic science methods in many animal species and other nonhuman species are highly underdeveloped. In the pigeon analyses, we have demonstrated that lineage analysis and

Table 1 Information Content of the 10 Pigeon Markers

Marker	No. alleles	PE	HTZ	PIC
PIGN 03	7	0.445	0.681	0.635
PIGN 04	9	0.587	0.762	0.742
PIGN 06	4	0.376	0.639	0.578
PIGN 10	9	0.557	0.748	0.723
PIGN 12	12	0.682	0.831	0.815
PIGN 15	8	0.504	0.728	0.687
PIGN 18	10	0.654	0.819	0.799
PIGN 26	8	0.703	0.852	0.834
PIGN 31	7	0.460	0.704	0.654
PIGN 57	5	0.463	0.707	0.659

identity verification is extremely useful in cases in which other evidence is circumstantial.

8.4 Identification of Stolen Cattle Using 22 Microsatellite Markers

L.H.P. van de Goor and W.A. van Haeringen Ph.D.

Several cattle have been stolen at a young age. To register these stolen calves, criminal farmers indicate that a twin has been born, to ensure that their own new-born calf as well as the stolen calf is accepted in the national database. This practice is quite difficult to identify because the twinning rate in cattle varies from farm to farm. A second method to register stolen cattle is to announce the birth of a calf when actually no calf has been born. Both methods are incidents of false reporting to the cattle registry.

In total, 35 cases of suspected stolen calves were analyzed. In all cases, we established DNA profiles for the suspected stolen calves as well as their likely dams and sires. One likely dam needed to be found on the farm on which a calf had been stolen. DNA was extracted from cattle hair roots using our laboratory's standard operating procedures. Multiplex PCR was used to amplify 22 genetic markers. Visualization of the DNA profile was performed using the ABI Prism 3100 DNA Sequencer (Foster City, CA). A computer program, CERVUS,[1,2] was used to compare the groups of cattle to each other.

A total of 141 cattle have been analyzed in the search for stolen cattle. Based on the standard set of 11 microsatellite markers, a large number of cases can be solved. In a limited number of cases, the markers lacked power to discriminate between possible dams originating from one farm. Consequently, we analyzed a second set of 11 additional markers on all cattle involved. This additional information allowed us to find matches between calf and dams in all cases.

As expected,[3] the initial testing based on 11 microsatellite markers did not provide sufficient power of exclusion to discriminate among cattle mated using line-bred principles. The addition of a relatively large number of microsatellite markers enabled the identification of the correct dam. It is highly important that cases such as the one we have described here are based on recommended forensic guidelines.[4] The cases presented here will certainly be taken into court, where judges will interpret the quality of the work. Since these kinds of cases are relatively new in the animal world, it is important to increase the minimum number of markers tested and desirable to generate allelic ladders as well as to incorporate duplicate testing of all samples involved. This will increase the level of confidence in the ability of DNA testing to correctly individualize highly inbred samples. Finally, in cases as described here, it is important to use as much information as possible. We have interpreted these cases mainly from a one-parent (dam) perspective, whereas the addition of genetic information from the sire would have improved the power of the analysis. At the same time, it is important to realize that on many farms, a limited number of bulls are used — sometimes as few as one bull for all calves produced.

References

1. Slate, J. et al., A retrospective assessment of the accuracy of the paternity inference program CERVUS, *Molec. Ecol.*, 9, 801–8, 2000.
2. Budowle, B. et al., Recommendations for animal DNA forensic and identity testing, *Int. J. Legal. Med.*, in press.
3. Sherman, G.B. et al., Impact of candidate sire number and sire relatedness on DNA polymorphism-based measures of exclusion probability and probability of unambiguous parentage, *Anim. Genet.*, 35, 220–226, 2004.
4. Marshall, T.C. et al., Statistical confidence for likelihood-based paternity inference in natural populations, *Molec. Ecol.*, 7, 639–655, 1998.

8.5 Development of a 17-Plex Microsatellite PCR Kit for the Genotyping of Horses

Pero Dimsoski, Ph.D.

In the past decade, the DNA forensic field was primarily concerned with the identification and cataloguing (DNA databasing) of the human genotypes. These efforts required development of sophisticated PCR-based genotyping technologies.[1–3] However, at the same time, a DNA genotyping technology (RFLP, STR, AFLP, SNPs) was developed for animals, plants, and microorganisms. Originally intended for human gene mapping, these new genotyping methods were also used to help with the classification of microorganisms and plants to the species level, animal and plant forensics cases, and to a large extent were utilized by animal and plant breeding societies for the construction and verification of large pedigree files and databases.

Modern tools of molecular biology provided means for fast, accurate, and relatively inexpensive ways for animal genotyping. Applied Biosystems' Stockmarks® line of products has been designed primarily for paternity testing and verification of pedigree. DNA genotyping as conducted by the kits is the most reliable method for parentage testing. The kits have also been used for tracking animal products to a particular herd, identification of animals in criminal cases, linkage mapping, and diagnostic studies. The ABI Horse Genotyping Kit[4,5] has been a great asset to the horse-breeding community, helping to genotype the pedigrees of tens of thousands of horses as well as to demarcate various horse breeds.[6] Considering that the price of the purebred horse can easily reach several thousands of dollars, the value of an accurate and efficient pedigree verification system by DNA genotyping can be put in rather significant economical perspective.

Genotyping laboratories have used the current Horse Genotyping Kit[4] primarily for pedigree and parentage verification, with the end-customers being various horse-breeding associations. The current users have been very satisfied with the kit performance; however, there was a need to update the kit to the new DNA technology standards. In the current kit all the primers are packed in separate tubes. Prior to PCR, the user had to mix the primers into two tubes and amplify the target loci separately. The kit consists of 12 primer sets amplified with two PCR reactions, an 8-plex and a 4-plex. The 8-plex reaction includes the following loci: VHL20, HTG4, AHT4, HMS7, HTG6, HMS6, HTG7, and HMS3; and the 4-plex reaction includes AHT5, ASB2, HTG10, and HMS2. One of the primers of each primer set is labeled with fluorescent dye that could be 5Fam, Joe, or Tamra. The products are separated on either the ABI 377 or ABI 310 genetic analyzer instruments and virtual filter set "A" is used to differentiate among the dyes' spectral compositions.

However, with the introduction of the additional fluorescent dyes and high-throughput instruments, it became necessary to update the kit in accordance with the new technology. For example, filter set "A" was no longer an option on the new generations of instruments (ABI 3100 and 3730); in addition, mixing the primers prior to PCR and having two PCR reactions was adding several extra steps to the laboratory procedures, resulting in unnecessary time and effort. The research and development goals were aimed to simplify the PCR protocols, combine all primer sets into one tube, implement new fluorescent dyes so the new kit could be used on the new instrument platforms, and, if possible, to develop new primer balances so all of the 12 primer sets could be amplify in one PCR reaction, possibly under the same cycling conditions utilized by the current Equine Paternity Kit. In addition, five other markers, already used by various genotyping laboratories, were included in the new kit, in order to increase the power of discrimination/inclusion.

The general procedures used for development of the horse kit are described by Wallin et al.[3] All primer sequences used in the kit are publicly available.[7] The kit loci, chromosome location, dye label, and the size range of the amplified products for all markers used in Horses Genotyping Kit are presented in Table 2. The following loci are included in the new StockMarks for Horses Equine Genotyping Kit: VHL20, HTG4, AHT4, HMS7, HTG6, AHT5, HMS6, HTG7, HMS3, AHT5, ASB2, HTG10, HMS2, ASB17, LEX3, HMS1, and CA425. The International Society of Animal Genetics[8] has recommended nine loci that are part of the kit. The remainder of the loci that were included in the kit were selected because various service laboratories in Europe and the United States utilized them. There is one primer pair per locus. One of the primers is labeled with fluorescent dye and the other is an unlabeled primer. There are four dyes for the primers: 6-FAM, VIC, NED, and PET. Also, LIZ dye is used for labeling of the size standard for use with the filter set G5.[9] The primers labeled with the 6-FAM dye are VHL20, HTG4, AHT4, and HMS7. Primers labeled with VIC are HTG6, AHT5, HMS6, ASB23, and ASB2, and

Table 2 Loci Names, Chromosome Location, Dye Label, and the Size Range of the Amplified Products for all Markers Used in 17-Plex Horse Genotyping Kit

Locus	Fluorescent dye	Chromosome Location	Size Range
VHL20	6-FAM	30	83–102
HTG4	6-FAM	9	116–137
AHT4	6-FAM	24	140–166
HMS7	6-FAM	1	167–186
HTG6	VIC	15	74–103
AHT5	VIC	8	126–147
HMS6	VIC	4	154–170
ASB23	VIC	3	176–212
ASB2	VIC	15	237–268
HTG10	NED	21	83–105
HTG7	NED	4	114–126
HMS3	NED	9	146–170
HMS2	NED	10	215–236
ASB17	PET	2	104–116
LEX3	PET	X	137–160
HMS1	PET	15	166–178
CA425	PET	28	224–247

Table 3 Thermocycling Condition for the Horse Kit for GeneAmp® PCR System 9700 Run in 9600 Emulation Mode

Temperature in °C	Time	Cycles
95	10 min	1
95	30 sec	
60	30 sec	30
72	1 min	
72	60 min	1
4		1

the primers labeled with NED dye are HTG10, HTG7, HMS3, and HMS2. The PET dye is attached to ASB17, LEX3, HMS1, and CA425.

The most optimal thermocycling conditions of the kit are described in Table 3. It is recommended to use the GeneAmp® PCR system 9700 in the 9600 emulation mode.

The above cycling conditions are exactly the same as the conditions of the current Equine Paternity Kit, which was one of the goals during the development of this new kit. The kit was optimized to be used with a total volume 15 μL PCR reaction. In addition to the primer mix, the kit will contain dNTPs, AmpliTaq Gold® Polymerase, and Stockmarks® buffer. The customer has to provide only PCR grade water. The amounts of each of the components for one PCR reaction are shown in Table 4.

The kit was optimized for use with a DNA template concentration ranging from 0.2–10 nanograms (ng). The electropherogram generated on a 3100 ABI Prism DNA Analyzer instrument by using template concentrations of 1.25 and 5 ng are presented in Figures 2 and 3, and Table 5, respectively. This kit was designed to perform well with low template concentrations (close to 1 ng) as is shown in Table 4 where peaks are above 200 relative fluorescent units (RFUs). The peak height and the color balance are good across loci for a DNA template concentration of 1.25, 2.5, and 5 ng. However, the 3100 instrument, because of its sensitivity, and because of the sensitivity of the kit, shows more pull-up peaks at the higher template concentrations (> 10 ng,

Table 4 PCR Components and the Amount for the Standard Reaction

PCR Component	Amount (μL)
Stockmarks® Buffer	2.5
dNTP mix	4.0
Amplification primer mix	4.0
AmpliTaq Gold® Polymerase	0.5
DNA Template	1.0
Deionized water	3.0

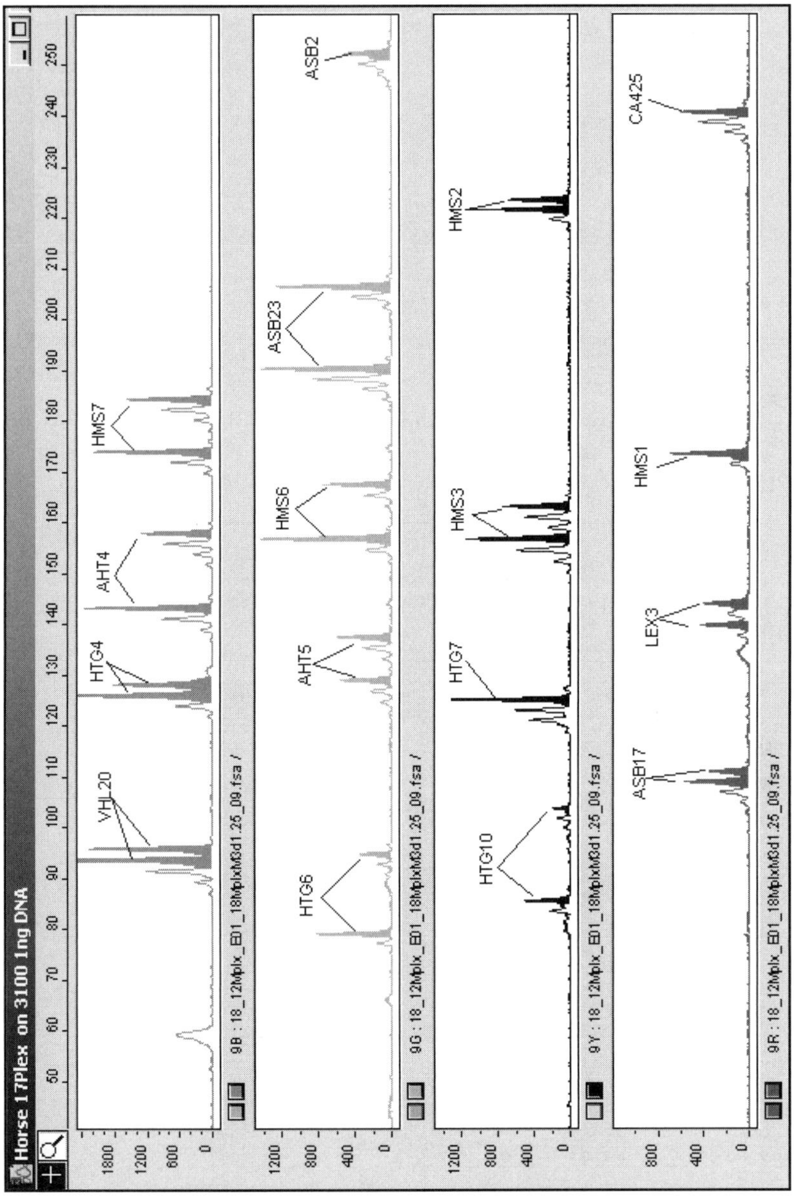

Figure 2 GeneScan electropherograms of amplified control horse DNA by 17-Plex Horse Genotyping kit run on ABI Prizm 3100 DNA Analyzer, [DNA] = 1.25 ng. (**See color insert after p. 108.**)

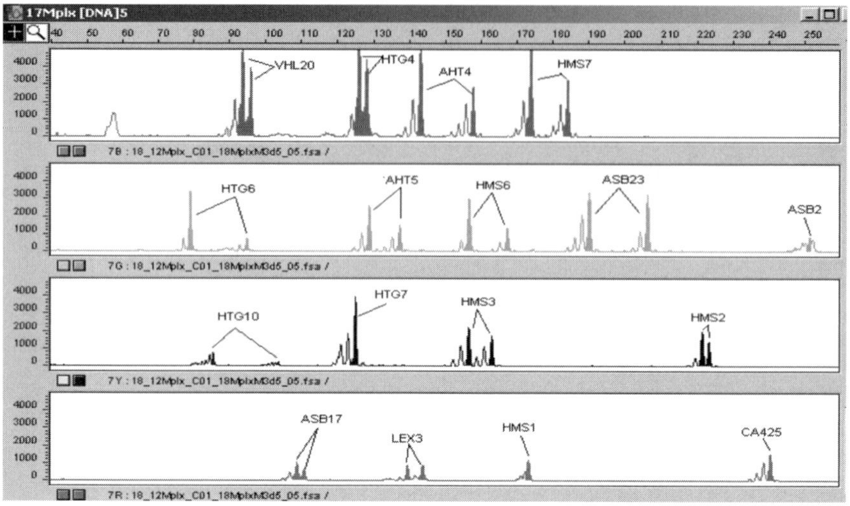

Figure 3 GeneScan electropherogram of amplified control horse DNA by 17-Plex Horse Genotyping kit run on ABI3100 instrument, [DNA] = 5 ng. (**See color insert after p. 108.**)

Table 5 Peak Heights, Run Time, Peak Area, and Data Point of Electropherograms from Figure 1

Dye/Sample Peak	Minutes	Size	Peak Height	Peak Area	Data Point
9B, 15	7.95	93.80	3520	37102	2981
9B, 16	8.02	95.95	2300	23687	3009
9B, 21	9.09	126.11	2852	33405	3409
9B, 22	9.16	128.22	1851	24668	3436
9B, 26	9.66	143.35	2371	22189	3621
9B, 31	10.11	157.98	1347	12425	3791
9B, 35	10.63	174.09	2206	22745	3988
9B, 40	10.97	184.38	1596	16636	4115
9G, 6	7.43	79.18	836	8076	2785
9G, 10	7.99	94.95	360	3762	2996
9G, 14	9.19	129.08	574	5906	3447
9G, 17	9.48	137.62	602	6128	3554
9G, 21	10.08	156.97	1445	15338	3779
9G, 25	10.42	167.67	759	7968	3909
9G, 32	11.17	190.43	1459	15159	4190
9G, 36	11.70	206.55	1288	13417	4389
9G, 41	13.16	252.58	519	7638	4935
9Y, 10	7.66	85.65	520	5290	2873
9Y, 14	8.30	103.77	201	1864	3113
9Y, 18	9.06	125.25	1346	12742	3398
9Y, 23	10.07	156.89	1157	11821	3778
9Y, 26	10.28	163.27	770	7947	3855
9Y, 29	12.19	221.72	866	8476	4572
9Y, 30	12.25	223.65	691	6864	4595
9R, 3	8.49	109.07	581	7524	3185
9R, 4	8.57	111.08	358	4647	3212
9R, 7	9.55	139.87	447	4928	3581
9R, 9	9.68	144.05	416	5983	3629
9R, 11	10.62	173.61	715	8754	3982
9R, 15	12.79	240.81	616	7157	4798

data not shown). Diluting the PCR products prior to loading them on this instrument can reduce the number of pull-up peaks. Since the sensitivity among the instruments varies, and the exact template concentration is not always known, there cannot be a strict recommendation about the dilution factors of the PCR products; instead, more general recommendations should be applied. For example, if the template DNA concentration is between 1 and 10 ng, then there should not be dilution of the PCR products prior to the detection run; however, if there are a lot of pull-up peaks and the template concentration is above 10 ng, then the dilution factor should be adjusted to the instrument sensitivity (as determined by each laboratory's validation). Therefore, depending on instrument sensitivity and the DNA template concentration, each user needs to adjust the running conditions of the kit. The new 17-Plex Horses Genotyping kit also performs well on the ABI 377 Prism instrument (Figure 4). In addition, the kit performs well if run on an ABI Prism 3100 with regular module (GeneScan36_POP4DefaultModule, Figure 5) and if run with a new module that enhances the color balance between different dyes (GeneScan36vb_POP4DefaultModule, Figure 6). The last module (GeneScan36vb_POP4DefaultModule) is the recommended module for Horses Genotyping Kit when used on ABI 3100 instrument. The kit has

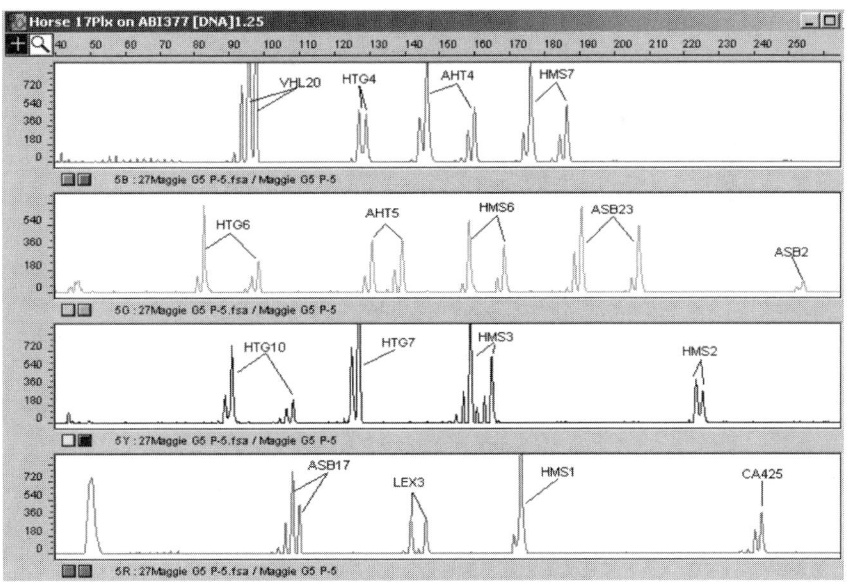

Figure 4 GeneScan electropherogram of amplified control horse. (**See color insert after p. 108.**) DNA by 17-Plex Horses Genotyping kit run on ABI377 instrument, [DNA] = 1.25 ng.

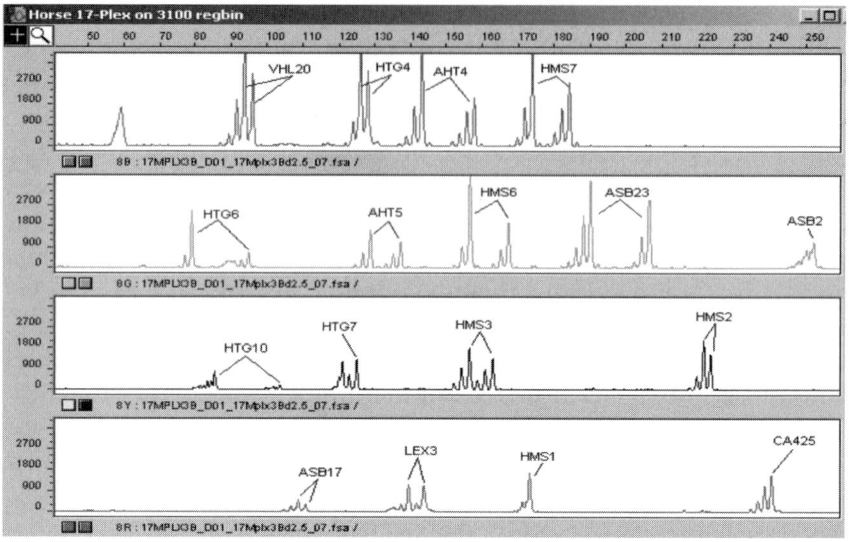

Figure 5 GeneScan electropherogram of amplified horse DNA by 17-Plex Horses Genotyping kit run on 3100, [DNA] = 2.5 ng, by using GeneScan36_POP4 Default-Module. (**See color insert after p. 108.**)

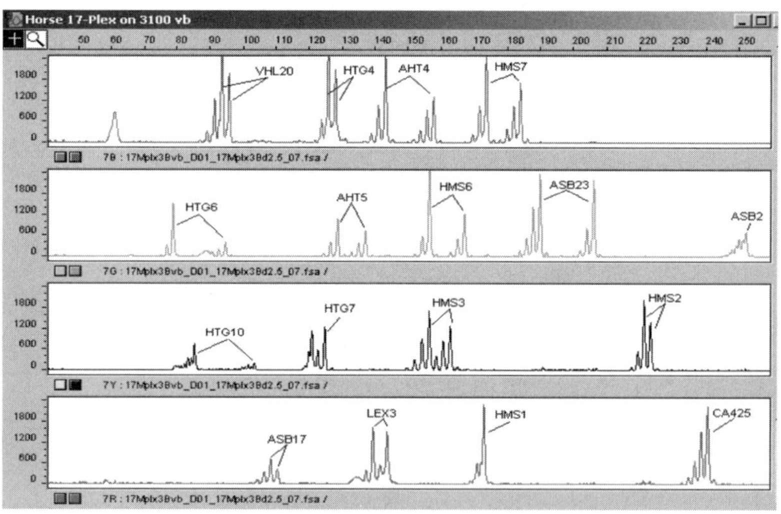

Figure 6 GeneScan electropherogram of amplified horse DNA by 17-Plex Horses Genotyping kit genotyped on 3100 run, [DNA] = 2.5 ng, by using GeneScan 36vb_POP4DefaultModule. (**See color insert after p. 108.**)

been tested in-house and off-site on horse DNA samples originating from various horse breeds. All of the primers amplified well with the exception of HTG10, which sometimes exhibited low peak heights that would not be suitable for automated scoring.

Breeding societies are primarily concerned with improvement and propagation of different breeds of livestock. A fast and accurate way to construct a pedigree is by knowing the genotype of parents and progeny. Therefore, there is a constant need to genotype all commercially available animals. In practice, horse breeders provide a parentage data to breeding societies, which enter the data into a registry and generate pedigrees. Even though this method worked well most of the time, it is not a very reliable way of pedigree verification, since it is prone to human errors at several levels (data collection, transfer of data, etc.). The most reliable and efficient method for pedigree construction and analyses is the one that employs the DNA genotyping technology. With the decrease in price of reagents and instruments, the DNA genotyping becomes the most cost-effective method for pedigree maintenance of large populations of animals. The 17-Plex Horses Genotyping kit has been designed to provide high discrimination power, with minimum time for sample preparation with minimum use of reagents. Therefore, the service laboratories who have high-volume contracts with the horse-breeding societies will be main users of the horse genotyping kit.

Another interesting development has been in the area of animal forensics. Over the past few years, the PCR-based method for genotyping of animals, particularly canine and bovine, have been used in high profile court cases.[10] Considering the number of pet animals in the United States, it should be expected that the field of animal forensics will be useful for animal tracking.[11] From these perspectives, the horse genotyping kit, described above, should be a welcome tool for forensic scientists and animal-theft investigators.

References

1. Wallin, J.M., Martin, M.R., Lazaruk, K.D., Fildes, N., Holt, C.L., and Walsh, P.S., TWGDAM validation of the AmpFlSTR blue PCR amplification kit for forensic casework analysis, *J. Forens. Sci.*, 43, 854–870, 1998.
2. Lazaruk, K.D., Walsh, P.S., Oaks, F., Gilbert, D., Rosenblum, B.B., Menchen, S., Scheibler, D., Wenz, M.H., Holt, C.L., and Wallin, J.M., Genotyping of forensic short tandem repeat (STR) systems based on sizing precision in a capillary electrophoresis instrument, *Electrophor.*, 19, 86–93, 1998.

3. Wallin, J.M., Holt, C.L., Lazaruk, K.D., Nguyen, T.H., and Walsh, P.S., Constructing universal multiplex PCR systems for comparative genotyping, *J. Forens. Sci.*, 47, 1–14, 2002.
4. Bozzini, M., Fantin, D., Ziegle, J., van Haeringen, H., Jacobs, W., Ketchum, M., Spencer, M., and Bates, S., Automated equine paternity testing, *Anim. Genet.*, 27, 32, 1996.
5. Marklund, S., Ellegren, H., Eriksson, S., Sandberg, K., and Andersson, L., Parentage testing and linkage analysis in the horse using a set of highly polymorphic microsatellites, *Anim. Genet.*, 25, 19–23, 1994.
6. Bjornstad, G. and Roed, K.H., Breed demarcation and potential for breed allocation of horses assessed by microsatellite markers, *Anim. Genet.*, 32, 59–65, 2000.
7. Horse map database available at: http://locus.jouy.inra.fr/
8. ISAG information available at: http://www.isag.org.uk/
9. Information available at: http://www.appliedbiosystems.com/support/software/310/modules.cfm
10. Information available at: http://www.animalforensics.com
11. Information available at: http://www.arc.agric.za/institutes/aii/main/divisions/animalbreedgen/animalgen/anfor1.htm

8.6 Pet Trading of Exotic Animals

Heather Miller Coyle, Ph.D.

While the pet trade and conservation biologists agree that parrots are threatened by habitat loss, they disagree about the effects of poaching. Avicultural interests downplay it but biologists say poaching chicks for the lucrative pet trade is one of the biggest reasons for the parrots' decline. New research shows that the biologists are right.

Nearly a third of the 145 parrot species in the neotropics (Mexico and Central and South America) are threatened, making them among the most endangered groups of birds worldwide. Parrots fetch an average of $800 in the United States and the number of parrot chicks taken from the wild is estimated at up to 800,000 per year. Parrots are particularly sensitive to poaching because they have low reproductive rates.

Many other animal species are subject to being removed from their natural habitat and sold either legally or illegally to pet shops for individuals who want an exotic pet. Examples of more exotic pets sold in local United States pet shops include the panther chameleon and several species of tortoise and fish (Figures 7 and 8).

Figure 7 Exotic pets like chameleons are difficult to care for but highly desired by pet stores and exotic animal lovers alike. (**See color insert after p. 108.**) Photograph is courtesy of John E. Coyle.

Figure 8 Tortoises are sold as pets but their shells and meat are also sold as delicacies in many Asian countries, leading to a steady decline in their population sizes. (**See color insert after p. 108.**) Photograph is courtesy of John E. Coyle.

Reference

For more information about the Society for Conservation Biology: http://conbio.net/scb/

Basics of Forensic Fungi

9

KHUDOOMA SAID AL NA'IMI, B.SC.

Contents

9.1 Introduction

Fungi are one of the largest and most heterogeneous eukaryotic families, with various species, forms, and a wide range of habitats such as air, land, and sea. The word *fungus* is thought to be derived from the Greek word *sphongis*, meaning sponge; however, others have suggested that the word could be derived from the Latin *Fingus ago*, meaning *I make a corpse.* The word *toadstool*, referring to a poisonous mushroom, comes from the German word *todesstuhl*, which means *death's stool.*[1]

The first fossil of a land plant and fungi appeared 480 to 460 million years ago.[2] It is estimated that there are approximately 1.5 million species of fungi; roughly 80,000 species of them have been described.[3] These creatures have a vital role in ecology and human life such as in fermentation, agriculture, food spoilage,[4] natural decomposition, and in various scientific and industrial applications. Sir Alexander Fleming, the discoverer of penicillin, said one day to a group of artists, "If any of you chaps has got a pair of moldy old shoes, I'd very much like to have'em." Fleming was curious about molds as part of his research of antibiotics.[5] In forensics, we may ask many questions about the growth of molds on a pair of shoes:

1. Which mold species are growing on those shoes?
2. How long did it take them to grow?
3. How large is the growth?
4. On which side is there more growth?
5. What are the effects of the mold growth on the leather?
6. Were the molds disturbed by some action during the crime?
7. What were the moisture and temperature conditions?
8. Has the mold digested any suspicious biological evidence?
9. Can the fungal metabolic products or actions be used forensically?

In the Sir Arthur Conan Doyle story, "The Musgrave Ritual," the body of Brunton the butler was discovered, and beside him was an open brass-bound wooden box, which was thought to contain the crown of Charles the First of England and coins. Doyle had Sherlock Holmes describe the box in the following way: "It was furred outside by a thick layer of dust, and damp and worms had eaten through the wood, so that a crop of livid fungi was growing on the inside of it."[6] Here the author connected fungi growth with dampness, but he gave little description about the mold conditions or whether the growth had been disturbed when most of the box contents were removed.

The study of fungi in forensics can be considered a sub–discipline of forensic botany, a novel and increasingly specialized discipline[7] with strong ties with other sciences such as microbiology, toxicology, and ecology. Fungi interactions with forensic investigation are many and include fungi growth and structures, pathology, taphonomy, decomposition, deterioration, biodegradation, mycotoxins, poisons, bioterrorism, wildlife forensics, and public health. The various connections and crossovers between several of these fungal utilities must be considered. This chapter is an attempt to give general ideas and to list some areas currently in use or areas that could be forensic research topics in the future.

9.2 Growth and Structures

Fungi can be unicellular (e.g., yeast) or multicellular (e.g., filamentous and mushroom). A filamentous fungi body consists of a thread of filamentous cells called *hyphae*, which have — in most species — connecting walls called *septa*;

the mass of hyphae is called *mycelium* (pl. *mycelia*). The reproduction method is either asexual or sexual by a fruiting structure (ascomata) that contains spores. Fungi are plants without chlorophyll and can grow in dry or wet (aquatic) conditions. Their major groups are zygomycetes, ascomyctes, basidiomycetes, and chytrids.[8] Generally. they can be a parasitic, saprophytic, or mycorrhizal form that is either lignophilic or coprophilic,[1] depending on the availability of energy–rich organic nutrients either from dead or living sources.[9]

Physiology and medium conditions control the growth rate of fungi, with some fungi being dimorphic. The germination of mycelial fungi starts under genetic control by spores growing a germ tube to produce hyphae by tip elongation and branching to form a mycelium. The hyphae produce a radial extension, which fuses by hyphal anastomoses when crossing each other, and grow into different directions. Most mycelial fungi have immortal growth and some species are used in aging research.[9] Fungi are also more tolerant of low pH levels.

The time scale of fungi growth could be very useful information in forensic investigations. For example, to approximate the last time food or beverage had been consumed, used, or prepared, one would want to know how long it took a colony or group of molds to grow on the item's surface. It has been noticed that the hairs (perithecial) on old fungal colonies of the genus *Chaetomium* — which prefers to grow in dark places — have a different structure than the hairs on young colonies;[10] this could be useful in forensic research to differentiate between the shapes of a new and old colony, and if any condition(s) stop them from growing, or if the present fungi species are correlated with the location and crime scene circumstances. One must keep in mind when estimating time of growth that numbers of blowflies such as *Calliphora* and *Lucilia* can distribute spores of stinkhorn fungus (*Phallus impudicus*), which attracts flies due to its rotting meat smells;, also, some blowflies are fungus breeders species.[11] Blowflies, which often provide important entomological evidence at crime scenes, have common parasitic fungi that can be the cause of their own death (e.g., *Empusa* and *Entomophthora*). These fungi can also cause some changes in beat in the wings due to the spore's action.[11]

9.3 Terrestrial Pathology

Fungi can cause health problems and consequently death as well as economic losses by infecting humans, animals, and plants. Generally, human infection can be cutaneous, subcutaneous, or systematic, and there are only a few species of yeasts, molds, and dimorphic fungi that can cause human infection. In the past, life–threatening infections due to true or opportunistic pathogenic fungi have increased; for example, invasive mycoses that are a growing public health problem can cause death in many immuno–compromised patients.[12]

Fungal infections can have forensic consideration; for example, their growth can be noticed in autopsy histological specimens, so an assumption that the possible cause of death is an autoimmune disease such as AIDS could be reached, in the absence of a medical history. Such responsible fungi are *Aspergillus* spp., *Fusarium* spp., and *Scedosporium* spp.[13] In one case, fungal cerebritis was the cause of death of two intravenous drug abusers. In the absence of any other identifiable predisposing factors, the microscopic examination showed a broad branching, non–septate fungal hyphae in one of the case tissue samples; there was no evaluation for cellular immune deficiency, but this could be the result of AIDS infection.[14] There are some concerns about the role of fungi and other organisms in Sudden Infant Death Syndrome (SIDS), because of their history of causing upper respiratory tract and ear infections, especially within the aboriginal populations in some countries, such as Native Americans, Eskimo, and Australian aborigines.[15] Another serious disease among infants is pulmonary bleeding or idiopathic pulmonary hemosiderosis (IPH) in infants, which is reported in homes with high levels of *Stachybotrys chartarum* fungi. This fungus produces proteins with hemolysin and proteinase activities that have been suggested to be a possible IPH factor.[16,17]

Some fungal infections on human bodies would not be recognized by police investigators who may think these are possibly marks of violence or torture (e.g., ringworm infection).[18] A pathologist or dermatologist would be able to identify these correctly. Medical laboratory workers could develop mycosis due to fungal infection by species such as *Coccidioides immitis* or *Histoplasma capsulatum*, if these incidents after full investigation appeared to have happened at work and are due to inadequate safety rules. In such instances, the infected employees could be entitled to compensation.[18] The fungal contamination of medical products or hospitals can also result in lawsuits if the contamination causes health damage or death, or if, after investigation, the contamination was proven to have been a trade, manufacturing, or services provider default. In a recent case, individuals who were wearing contact lenses in the United States were advised not to use a specific contact lens solution (ReNu with MoistureLoc) after the solution was associated with a multi–state outbreak of *Fusarium*, which can cause blindness and require corneal transplantation.[19] The manufacturers could suffer economic losses internationally in such civil action cases. Other fungal species such as *Aspergillus*, *Penicillium*, and *Candida* can also cause Keratitis.[20]

9.4 Mycotoxin

Filamentous fungi can produce mycotoxin substances as secondary metabolic products. They are natural chemical components and considered to be toxic for animal and other vertebrates in low concentrations. Some fungi species can produce one or more toxins, such as *Asperigillus flavus*, *A. ochraceus*, *A. terreus*, *A. parasiticus*, and *Penicillium expansum.*[21] About 200 fungi species can produce an estimated 300 different mycotoxins.[22] Examples of these toxins are aflatoxins (G1, B1, and M1), citrinin, fumonisins, ochratoxin (A, B, and C), patulin, trichothecenes, and zearalenone. Some of these mycotoxins can have combined effects.[23,24]

Mycotoxins may require forensic investigation, as they have economic and health importance. In agriculture, mycotoxin contamination can occur in some products either naturally or due to bad storage conditions. Products that can be contaminated include almonds, nuts, chilies, corn, peanut, pistachios, wheat, and soy beans. The mycotoxin can spread to humans either directly or through animals via the food cycle. Some mycotoxins are considered more important than others due to their high level of occurrence and their toxicity to humans and animals. Many countries have approved legislation to set acceptable limits for mycotoxin levels in food. Such limits are necessary in order to reduce the intake of mycotoxins and protect public health.[22] As a sign of forensic importance, in some instances, United States experts have been called into court to testify on the mycotoxin contamination of stored grain cases.[25]

Ochratoxin A (OTA) has been reported in high frequency, but low concentration, in the blood samples of individuals in some countries. The composition of fungal populations and their mycotoxins in agricultural products will depend on the conditions and length of storage. OTA can be present in wines,[21] depending on the contamination level of the grapes either before or after harvest.[26] In the United Kingdom in early 1960, more than 100,000 turkeys died due to the consumption of food contaminated by aflatoxin B1, which has been called *Turkey-X Disease.* Research shows that aflatoxin B1 also has a carcinogenic effect on animals, and the World Health Organization (WHO) has classified it as a carcinogenic agent to humans.[22]

There are also concerns that the mycotoxin ochratoxin A, which is found in high levels in the natural environments of several Balkan countries (e.g., Bosnia, Croatia, Serbia, Bulgaria, and Romania), could be the etiological agent of the Balkan endemic nephropathy (BEN). In some of these Balkan countries, the level of this toxin was found to be high in wheat and corn samples.[23] In Africa, there were 317 cases of acute aflatoxicosis due to the consumption of maize with high levels of aflatoxin. The consumption resulted in death for 125 (37%) of those who consumed the maize.[27] In 1930, the death of livestock in Russia — mainly horses — was blamed on the presence of the fungus *Stachybotrys chartarum,* which contaminated the animals' food, such as hay. The animals were very sensitive to these fungal toxins, which have been recognized in later studies as satratoxin H.[16] In the United States, fungal endophytes of grasses cause livestock disorders and death, and the annual losses in the beef industry in 1990 was estimated to exceed $600 million due to the lethal and sublethal effects of these fungal toxins.[28]

9.5 Mushroom Poisons

Eating wild mushrooms (Order *Agaricales*) is an international cuisine tradition. Picking or hunting mushrooms is popular among many people; in the Czech Republic 72% of families pick mushrooms for their household consumption.[29] Edible fungi are a safe species for consumption and use, they are rich in proteins and essential amino acids, and they improve the immune system. In China, medicinal fungi have been well known for over 2,000 years. Examples of edible fungi are button mushroom (*Agaricus bisporus*), morels (*Morchella* spp.), shiitakes (*Lentinula edodes*), split-gill (*Schizophyllum commune*), and baker's yeast (*Saccharomyces cerevisiae*).[30]

It is estimated that there are over 5,000 species of mushroom worldwide, whereas 50–100 species have been reported to be poisonous[1] (e.g., amanita mushrooms, which comprise 33 or more species[31]). Mushrooms also can contain a high concentration of environmental toxic elements such as mercury, which is effectively absorbed by some species from polluted soil.[29] The causes of human poisoning by consumption of mushrooms can be due to mistakes, misidentification, or eating curiosity either by adult or children; and possible criminal acts must be considered in some cases. Individuals react differently to mushroom poisons[32] and the poisons can be divided according to their principle toxin groups: cyclopeptides, orellanine, monomethylhydrazine, disulfiram-like, muscarinic, isoxazoles, and gastrointestinal irritants. In untreated cases of amanita mushroom poisoning (*Amanita phalloide*), the mortality rate is between 30% and 90% and it can also cause liver damage.[1] In Japan, from 1991 to 2000 there were 947 reported cases of mushroom poisoning. Most of the cases were not analyzed toxicologically because the majority of them were caused by *Rhodophyllus rhodopolius* and *Lampteromyces japonicus,* which cause less serious forms of mushroom poisoning.[33]

Poisoning by ingestion of fungi can be a significant health issue. A 75-year-old man and his wife consumed mushrooms in their lunch and dinner in Japan. The next morning, the wife was found dead and her husband walked to the hospital for help. The forensic investigation recovered a piece of mushroom from the trap of the sink, which was analyzed, in addition to the wife's blood, brain, and stomach contents. The results revealed the presence of the amanita poison Alpha-amanitin in all the samples verifying the cause of death from the consumption of poisonous mushrooms.[33] In Portugal, a poisoning case of *Amanita phalloides* (death cup) mushroom was reported when four people were admitted to the hospital after eating wild mushrooms collected near their home in a suburban area. The first patient was a 30-year-old woman, admitted 25 hours postingestion; she recovered after medical treatment. The second patient was the partner of the first patient, a 24-year-old male, who was admitted 46 hours after ingesting a large portion of the mushroom, which caused him serious liver damage that required an orthotropic liver transplant. The third patient was the 12-year-old son of first patient, admitted 12 hours after ingesting a very small portion of the mushroom; he recovered after medical treatment. The fourth patient was a 2-year-old female, the child of a neighbor, who was admitted 12 hours after ingesting the mushroom; she underwent orthotropic liver transplantation due to liver damage.[34] In a case from Turkey, a 2.5-year-old girl consumed two mushrooms her mother had collected from the woods as food believing the mushrooms were a good source of protein. After 10 hours, the girl developed vomiting, diarrhea, and later hepatic encephalopathy, a condition consistent with severe brain edema, and possible brain stem herniation. She died after four days and 16 hours. At autopsy, her liver was found to have massive hepatocellular necrosis and macrovesicular steatosis. Generally, children are more affected by the poisons due to their smaller body size.[35]

9.6 Illicit Mushrooms

Hallucinogenic fungi, also called *magic mushroom* or *shrooms*, are starting to be a social problem in many countries, such as the United States,[36] Germany,[37] Canada, and Japan. These types of fungi can be found growing naturally in the wild or can be cultivated in gardens, inside homes, or bought illicitly as kits for cultivation. The kits consist of mycelium on a growing medium, spores, or dry mushroom. These species (*Psilocybe cubensis* and *Copelandia* genus[38] contain the controlled substances psilocin and psilocybin, which have psychoactive effects.

In Japan, a 27-year-old man was found dead in an irrigation canal. During the forensic investigation, two cultivation pots of mushrooms were found in the deceased residence. The pots were placed in incubation in the laboratory to cultivate the mushroom until its growth completed after two weeks. The mushrooms were identified from the spore as the hallucinogenic fungi species *Psilocybe subcubensis*. Analysis of samples from both the mushroom pots and postmortem blood showed the presence of hallucinogenic substances. The cause of death was reported as the result of cold water in the canal (hypothermia) while under the influence of the hallucinogenic substances he produced at his home.[33] In another case from Poland, a group of five young persons used a mushroom key book to gather six fruiting bodies of fly agaric mushroom, (*Amanita muscaria*). They skinned, shredded, and dried the mushroom in an oven and later consumed it with bread, salad, and beer. After 20 minutes they experienced auditory and visual hallucinations. One girl from the group suffered a disturbed physiological reaction that required medical intervention and recovered. There was no forensic investigation mentioned due to the notoriety of the individuals of these cases.[39]

9.7 Indoor Toxic Mold

Fungi grow in damp and moist places such as homes, offices, work areas, or schools;[40] they can cause health, economic, and safety damage, which can raise insurance claims and compensations. These damages can include weakness of building structures, unpleasant conditions (bad smells and appearances), and possible negative health effects from mold toxins. Sick building syndrome (SBS) is caused in part by mold,[41] especially in water-damaged buildings.[42] Fungal growth in these buildings can release spores and other smaller particles that could contain inflammatory, allergenic, and mycotoxinlike reactions.[43] Air in places with a significant amount of mold can be a real hazard due to possible effects of airborne *Stachybotrys chartarum* toxins such as the macrocyclic trichothecenes mycotoxin.[44]

Flooding in New Orleans (USA) caused by hurricanes Katrina and Rita in August and September in 2005 resulted in approximately 40,000 homes (17%) with heavy mold contamination.[45] This could be a huge health problem and could lead to extensive home damage insurance and investigation claims. In the United States, a bill was introduced to establish legislative grounds for indoor toxic mold industries; the bill included research, inspection, remediation, and insurance standards[46] and also could produce lawsuits against these industries. In particular, military personnel on international missions could suffer the effects of unanticipated living conditions that support mold growth. Mold exposure could result in health problems that would likely reduce their effectiveness in their duties. In a study on Finnish peacekeepers in Kosovo that evaluated different risk factors encountered in their daily duties, water-damaged buildings and mold risk represented 30% of most perceived risks according to the collected soldier's interviews.[47]

9.8 Decomposition and Taphonomy

Fungi are one of the major factors aiding decomposition on earth. Fungi have been reported on decomposed plants, humans, and other animal bodies and remains, either in the early stages of decomposition or later, as in the case of humans.[48] A few researchers have suggested using some fungi species as clandestine grave markers and in the estimation of postburial interval (e.g., *Hebeloma syrjense* in North America).[49] There are some disputes when considering some species of fungi as standard species of postputrefaction[50] where geographic location and environmental conditions must be taken into account; more study is needed in this area of fungal classification.

It is clear from research studies that when insects and other biological decomposition agents are not permitted to reach the organically rich dead bodies/cadavers, the only organisms that can lead to decomposition are fungi and bacteria when the conditions are suitable for them to grow. Fungal hyphae growth on biological specimens can make changes to the body due to the enzymatic secretion and absorption from the fungi cells. For example, Figure 1 shows how a flesh fly head has been deformed by fungi growth, after keeping the fly inverted in a tube in a refrigerator at 4°C for a few months.

Fungal growth on a cadaver found outdoors can provide useful evidence to estimate the postmortem interval (PMI). Using this technique, a group of researchers from Japan noticed a white spotted fungal growth on the external face of a cadaver removed from a water well. They sampled and cultured the growth and identified the fungi to be *Penicillium* spp. and *Aspergillus terrous*. Taking into account the previous research on fungal colonization, environmental conditions, state of decomposition of various organs, and police information, the investigators were able to estimate PMI two days shorter than the last time the deceased had been seen alive.[51] Fungal growth on human remains found in the forests (e.g., mummified cadavers) have been identified with the species of *Eurotium repens* and *Eurotium rubrum*, and

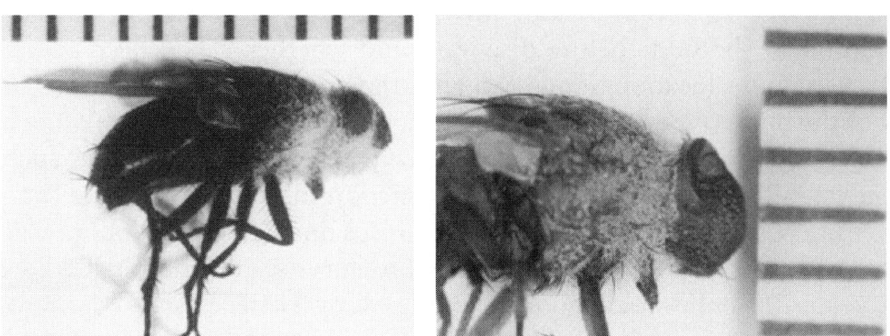

(a) (b)

Figure 1 (a) Fungi growth on flesh fly body, (b) A deformed head of the same insect after a few months of infection (scale: 1 mm).

from skeletal remains with the fungus species of *E. repens*, *Eurotium chevalieri*, and *Gliocladium* spp. It is quite typical for forensic pathologists to encounter fungi during their work.[48]

Researchers from the University of Al Anbar, Iraq studied fungal growth on cadavers fixed in formalin used in the medical school for teaching purposes. By taking swabs from the exposed cadaver and from the formalin pool with suspected fungal colonies, and doing the work of culturing and examination in the same storage room of the cadavers, they identified five different fungal genera and recorded their development time in the media as follows: (1) *Aspergilli* (2–3 days), which include also *Aspergillus fumigatus* and *Aspergillus niger*, (2) *Penicillium* (2–3 days), (3) *Trichophyton* (after 10 days), (4) *Epidermophyton* (after 7 days), and (5) *Cryptococcus* (rapid growth in 2 days).[52]

Professional centers such as forensic anthropology labs, osteological museums, morgues, and dissection and cadaver storage rooms can develop fungal growth on their collections. Fungi spores are easily spread and can develop in these places due to the moist conditions or problems in chemical fixation or cooling.[52] The growth in the case of bone tends to start on the soft ends as shown in Figure 2, which could develop and be a source of contamination of other bones. Some possible moisture sources are from the field environment, water maceration processes,[53] laboratory washing without enough drying because of cold weather, equipment, or human error due to work place pressure or lack of internal procedures for quality control. In addition, storage conditions of moist packaging or room water leakage promote fungal growth. Addressing these problems and reducing the likelihood of their occurrence is an ethical, scientific, health, and safety issue. The use of desiccants or dehumidifiers could be useful in some situations.

Understanding the effects of fungi on different parts of a human cadaver — such as bone and hair — can provide valuable information about the condition of the body before discovery and whether it had been removed from a previous location. Fungi can alter the microstructures and morphology of the bone (Figure 3) by hyphae penetrating the bone's hard tissues and by leaching minerals from it through enzymatic action.[54] Microbes such as fungi can also produce histological destruction in bone by forming tunnels either internally or on the bone surface. This is one factor that does not allow most buried bone in normal conditions to survive into the fossil record.[55] The genus *Mucor* has been reported to cause fungal osteoclasia — the absorption and destruction of bone tissue — in buried bones. The study shows that the bone material could be absorbed by the fungus membrane after solubilizing the inorganic bone fraction.[56]

Hair evidence is very important in a crime scene. Some house dust fungi such as *Chrysosporium indicum* can complete hair deterioration in 24 days.[57] The hyphae of fungi tend to tunnel in the human hair by penetrating in any location; the opening appears as a dark spot under an electron microscope.

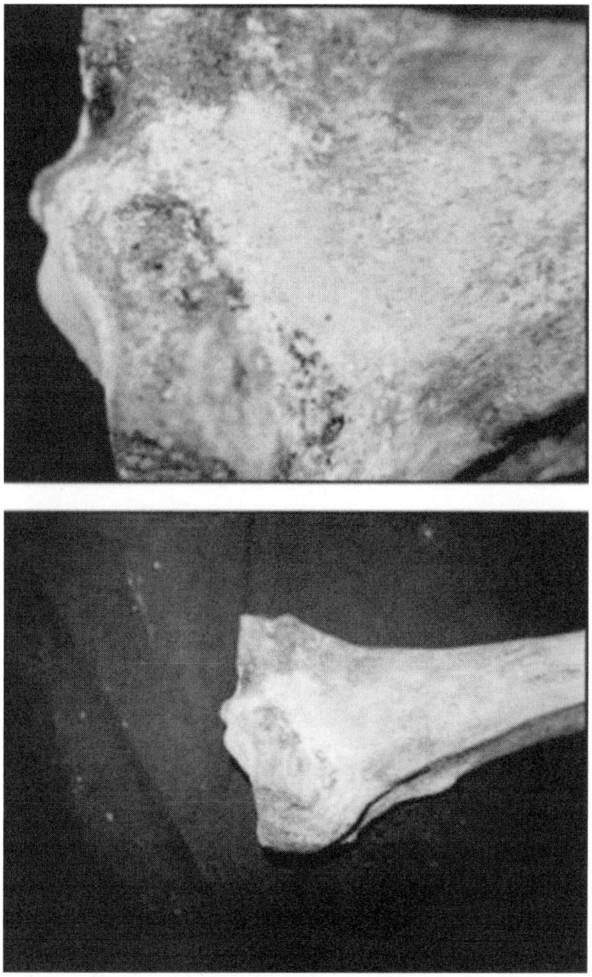

Figure 2 Fungus growth on human tibia, proximal end, posteriorly. (**See color insert after p. 108.**) (Courtesy of International Commission on Missing Persons, Bosnia and Herzegovina)

The tunnels are of small diameter with minimal branching,[58] and can reduce or destroy the hair's genetic or morphological contents.

Most of the published research on fungal decomposition is on plants and other crop species. The saprotrophic fungi undergo a succession (replace one another) in space and time on dead plant material.[59] Wood-decaying fungi have been divided into three types: white rot, which is active on cellulose, hemicellulose, and lignin components; brown rot, which is active on polysaccharide components; and soft rot, which is active also on polysaccharide and less so on lignin.[60] This research could be useful for studying forensic evidence with similar conditions.

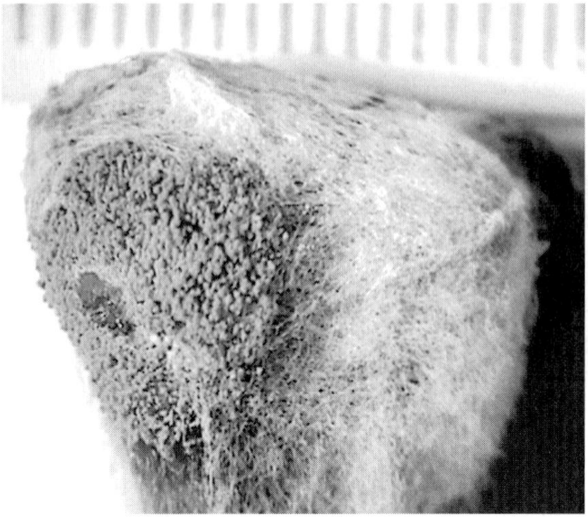

(a)

(b)

Figure 3 (a) Fungus growth on chicken bone (scale: 1 mm) kept in a closed container with some moisture at room temperature for few weeks, (b) a close-up photo of some fungal fruiting bodies. (**See color insert after p. 108.**)

There is limited literature on animal decomposition and the relationship between fungi and weather, chemical, and biological factors in terrestrial, freshwater, or marine habitats. Putting a fresh animal liver in a closed container at room temperature will produce a fungal growth, which will start to digest the tissue and produce spores in later stages as the structures and colors of fungi colonies change (Figure 4). In forensic science, there are "body

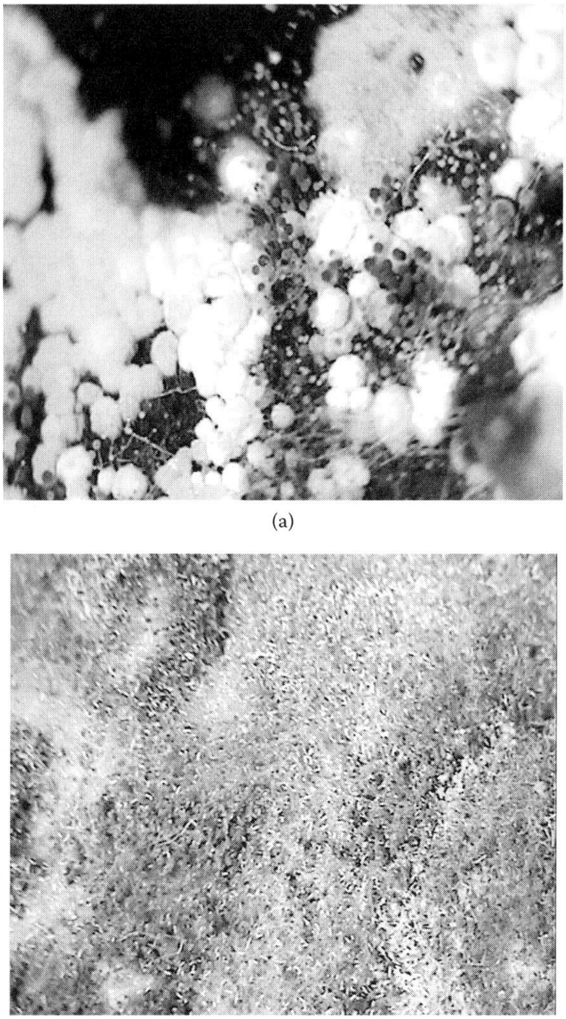

(a)

(b)

Figure 4 Fungi growth on animal liver tissue: (a) after several days, and (b) after a few months. (**See color insert after p. 108.**) The sample was kept inside a closed plastic container at room temperature.

farms" where experimental decomposition is studied, but more work needs to be done in this area of research.

The interaction of fungi with animal products in the food industry is widely known, and this could be a useful area in forensic research for civil cases or for food regulatory agencies. For example, fungi on dry, cured meat occur after several months if the food items are not stored properly. Such meat products include sausages, ham, and pork loins. Fungi growth, particularly on the outer muscle tissue, acts by digestion of the myofibrillar proteins by proteolytic enzymes, which will produce free amino acids that contribute to meat flavor and will be degraded to volatile compounds.[61–62] As a side note of fungal research interest, bacteria and fungi are also present in space, and their sources could be space shuttle material, air, water, food, or astronauts' bodies.[63] The microbial effect on an astronaut's health in space is well known and documented.[64]

9.9 Hydrocarbon and Fungi

Fungi can degrade and synthesize hydrocarbons. Some of these compounds are essential in forensic investigations, especially in arson investigations, and in the false positive results of alcohol increase in dead bodies and postmortem blood, detected by modern analyzing systems. Hydrocarbon detection is very important in arson investigations to determine whether a hydrocarbon accelerant was used to set the fire. Fungi and other microbes can degrade these compounds beyond forensic identification in specific conditions. To prevent degradation, a nonvolatile bactericide can be added to arson residues, or the samples can be refrigerated until the analysis time. In addition, doing a test on soil to ensure it does not contain microbial species capable of degradation of hydrocarbons could be useful.[65] Such fungal species are *Aspergillus*, *Penicillium*, and *Rhizopus*, which could also be used in oil pollution control.[66]

Fungi also synthesize and produce forensically important hydrocarbons, such as ethanol,[67] which is an indication of alcohol consumption. These organisms, in addition to bacteria, can raise the ethanol level in postmortem bodies.[68] *Candida albicans* is a form of yeastlike fungi reported to increase or produce ethanol in postmortem human stored blood. As a result, it is recommended that sodium fluoride be added to blood samples to prevent *C. albicans* and other microbes from growth.[69] Fungi also can produce volatile hydrocarbon compounds such as toluene and xylene from the fungal strain of *Penicillium roqueforti*.[70] These can give false positive results in analysis; in addition, they can produce a controversial interpretation either in the legal context or for the family of the deceased.

9.10 Bioweapons

In the modern era, the possible use of biological agents and toxins as war weapons has decreased. In contrast, the chance of bioterrorism or bio-criminal use has increased. The serious threat posed by biological weapons is a result of the varied consequences and challenges they can produce in the medical, psychological, political, scientific, and economic levels in society, taking into account the size, type, and directions of the attack.[71] Generally, fungi used as a biological agent in war or in criminal planning can be in the form of toxins, poisons, or living agents. The use depends on the perpetrators' target, which may be either humans or agricultural or food storage places. For example, Trichothecene mycotoxins can be used as a weapon and delivered through mixing with food or beverages or dispersed in air, which will produce serious dermal, oral, ocular, or respiratory medical symptoms that can lead to death.[72]

Some of the potential fungal biological terror agents for humans and animals are *Coccidiodes immitis*[73] (the agent for valley fever), *Histoplasma capsulatum*, and *Nocardia asteroidean*.[74] The fungus *Coccidiodes immitis* is known to be endemic in countries such as the United States, Mexico, Venezuela, Bolivia, Guatemala, Honduras, Colombia, Paraguay, and Argentina. *Histoplasma capsulatum* has been identified in at least 60 countries, and can be present in soil and environmental material.[75,18]

Historically, the use of fungi or their products as harmful agents has been reported in several cases. For example, following World War II, the Soviets used *Fusarium* to contaminate the food of Russian civilians.[71] A toxic attack occurred in 1982 against the Khmer Rouge guerillas at Tuol Chrey, in Kampuchea, after an artillery shell exploded upwind of the camp. The soldiers detected a sweet perfumelike odor and suffered symptoms of intoxication. There were at least 100 casualties and one death. The causative agent was a combination of trichothecenes mycotoxins such as T2 and HT-2.[17,76]

Fungi can also be used as a living infectious agent on crops in order to produce economic losses and to threaten food security. Among the highest-ranked of the fungi pathogens and toxins listed in the USDA are T-2 toxin, *Ralstonia solanacearum* (brown rot), *Synchytrium endobioticum* (potato wart), and *Phakopsora pachyrhizi* (soybean rust).[71]

Fungal bioweapons can also be used in a positive manner, known as biological control. Some fungal species have been reported to be pathogenic of opium poppy (*Papaver somniferum* L.). They can be used as biological agents to control plant cultivation in drug control and prevention programs.[77] The virulent fungus *Pleospora papaveracea* was developed by the UK Foreign Office to target and destroy the main varieties of Afghan opium poppies in order to eradicate the farming of this source of heroin, an industry with an estimated worth of £15 billion annually.[78] The method for developing this species as a biological control agent could also be used to develop other fungal species as bioweapons agents designed to destroy other plants of economic significance.

9.11 Palynology

Forensic palynology is the science of using pollens, spores, and other acid-resistant micro-plant remains (modern or fossil) in a legal context, which has many applications.[79,80] Fungi spores may be relatively ubiquitous or geographically restricted and their sizes differ between species. For example, the fungus *Psilocybe cubensis* spores have a length of 15.30 μm, and width of 9.28 μm, while *Fusarium moniliforme* spores have a length of 7.84 μm and width of 1.81 μm.[81] Spore development time and concentration also can differ between some species.[43] The possibility of using information about the different development times for individual species could be useful in a forensic investigation to know the time needed for fungal spores to form in a specific place and under specific conditions, and whether any conditions prevent the spores from forming, to connect it to the forensic investigation.

In one forensic case investigation, a young girl was sexually assaulted outside. Investigators collected plant pollen grains from the supposed crime scene, victim's clothes, and vaginal swabs. The investigators determined that the pollen grains from each source were identical. The unusual characteristics of these pollen grains helped to confirm the sexual assault location. The pollen grains had a fungal hyphae growth inside them, due to the site's dampness, which supported this type of fungal-pollen association.[80] Moreover, fungi may also deteriorate pollen grains, which should be considered when examining pollen evidence.

9.12 Aquatic Fungi

There are many species of fungi growing in freshwater or marine environments that could be useful in a forensic investigation. For example, some species have been reported to grow on submerged dead plant fragments.[82] Zoosporic aquatic fungi have been reported to grow on dead zooplankton organisms of freshwater.[83] One of the waterborne fungal infections caused by the genus *Batrachochytrium dendrobatidis* (order Chytridiales) is the agent for the amphibian chytridiomycosis infectious disease, which is responsible for the massive death in amphibians (frogs) in many countries such as Australia, United States, and Central and South America.[84,85] This information could be useful in wildlife forensic investigation.

Fungi also can develop in biological specimens stored in an aquatic medium. For example, the author put an insect (wasp) in a closed container filled with tap water and kept it at room temperature. After one week, a white fungal growth was over the insect's body, with motile spores swimming in the water (Figure 5).

Marine fungi are widespread in sea water and more than 500 species have been described.[86] They could be potential targets for research in cases such as drowning, decomposition, illustrating contact in a marine area, material damages, and seafood poisoning. Not all mold are considered to be fungi; some of them, such as water mold (oomycetes), which can grow on dead or live surfaces, has been reclassified from fungi to the protozoa group in modern classification. Oomycetes can cause ulcerative lesions in fish, where hyphae will either grow

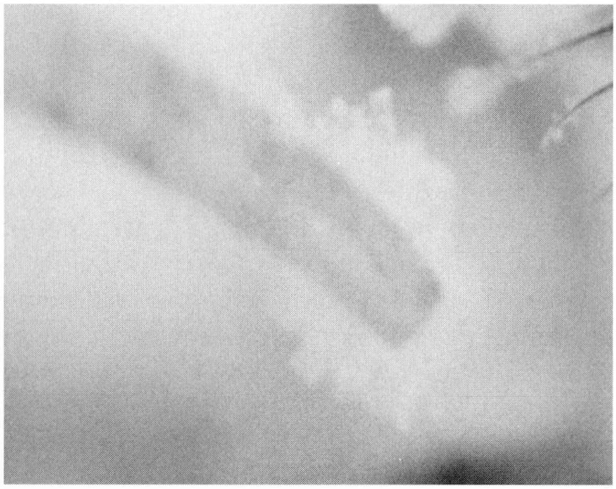

Figure 5 Water mold growth on the tip of an insect antenna (wasp) immersed in water.

on surface dermis or form deep focal chronic ulcers that appear as wounds on the fish body.[87] Some species of marine fungi also could represent a real risk of poisoning through the consumption of contaminated shellfish, which accumulate toxic metabolites of these fungi via filter-feeding.[88,89]

9.13 Deterioration and Biodegradation

In the beginning of my career in forensic science, investigators brought some clothes from a crime scene. The clothing had been partially buried in a desert area. I noticed there was some fungal hyphae on them and that the color of the material was a little lighter in that area. I asked whether these observations could be used to understand the case condition, but I did not find a clear answer at that time. Fungi are universal decaying agents. They have the ability to deteriorate and biodegrade materials such as dyes, textile, leather, hydrocarbon fuel, cinematographic films, papers, wood, concrete, old documents, stone, plaster, and cement[64,90] due to their enzymatic and physiological abilities.

Textiles such as clothing and fibers of cotton, silk, linen, hemp, jute, and wool are important clues at a crime scene and their deterioration and character can help investigators understand additional details about the crime. The rate of deterioration of textiles depends on factors such as biological agents, chemicals, temperature, humidity, moisture content, light intensity, reactivity, and nature of the material.[91] The characteristics of the textile could also reveal the condition in which it was before discovery, such as the burial environment by studying the fungal activities.[49] For instance, 34 fungi from 13 different genera were tested for their ability to degrade the protein keratin, which form 99% of wool fiber. Wool degradation by *Aspergillus* and *Penicillium* spp. will pass in 3 stages, first surface structure erosion, second swelling and production of spindle-shaped cortical cells bundles, and third cortical cells separation (97) (Janaway, 2002).

Some species of fungi have the ability to change the color of dyes (biological dye bleaching) by invading textiles or other colored items, discoloring the fibers, and alterating their dyeing properties.[92] Fungi such as white-rot fungi (WRF) are starting to be used in the dye industry as enzymatical biodegradators for toxic dyes in water.[93,94] The fungi's mechanism of decolorization is either dye adsorption to the fungal mycelium or degradation of the dye molecule by the process of oxidation. The enzymatic system of such fungi includes lignin peroxidase (LiP), manganese peroxidase (MnP), and laccase.[95]

Evidence made with leather, such as wallets, shoes, bags, belts, and jackets, can be subjected to moisture at burial sites (a complex set of conditions) or other places, which will enable fungi and other microbes to damage them. This can change an item's characteristics, as the leather is a source of nutrition.[96] The growth depends on factors such as the moisture level, leather type, and pretreatment of the leather product;[97] for this reason, desiccants are used in most stored leather products.

More than 30 species of fungi such as *Fusarium*, *Aspergillus fischeri,* and *Cladosporim resina* have been reported to grow on military fuel and their distribution system, which can cause corrosion in the fuel tubes and storage tanks, block pipelines and filters, and reduce the quality of the fuel.[98] Fungi have been also detected in diesel fuel tanks in Brazil.[99]

Filamentous fungi can deteriorate cinematographic films by hydrolyzing their gelatinous components if they are stored under inappropriate temperature and humidity conditions. This can affect the contents and characteristics of these films.[100] In high relative humidity (RH) and temperature, electro-technical insulation materials can be attacked by fungi that could cause biocorrosion and equipment failure.[101] Fungi also can cause biodeterioration of polymeric materials, affecting a wide range of industries, such as the space program. Technophilic fungi and bacteria caused metal and polymer degradation in the Russian MIR space station.[63]

Other surfaces, such as archaeological sites, historical monuments,[60,102] wood bridges, private and public buildings, other artifacts[103] and damp building materials,[104] can be colonized and damaged by fungi bio-films. In the case of a deteriorating and decaying concrete building, fungi can cause the bio-solubilization of several metabolic elements, for example, acidification of calcium; examples of these fungi are *Aspergillus ochraceus* and *Aspergillus niger.*[102] Knowledge of this process can be useful in investigations of archaeological sites and of building destruction and deterioration in forensic engineering. Fungi are ubiquitous and can grow on almost any surface, for example, on the arms of a watch if water leakage occurs, as happened to

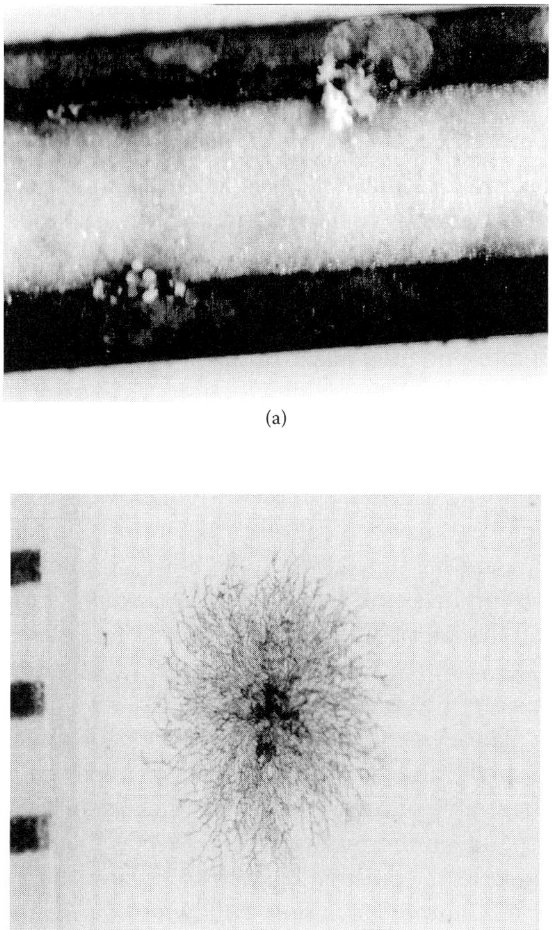

(a)

(b)

Figure 6 (a) Fungi growth on a wristwatch arm surface and (b) Fungus colony growth on a plastic bag surface (scale: 1 mm).

the author during his visit to Bosnia and Herzegovina, or on the surface of a plastic bag under suitable conditions (Figure 6).

9.14 Investigation Methods

Fungi evidence can be important in either a civil or criminal investigation, and as a rule the worker must follow the strict procedures governing the forensic practices of investigation, collection, and preservation. The suggestions made by Coyle et al.[105] are very useful and should be considered in general forensic botany investigations, especially in evidence recognition, documentation, collection, and preservation. The evidence chain of custody is important in forensic work; samples that will be used in legal cases must be documented from the time of collection to the final stages of analysis and logging. To protect the integrity of the evidence and the forensic test results, the following information may be required: the involved persons' names, description, collection location, date, time, identification numbers, barcodes, transport condition, laboratory name, and analysis methods.[41]

As a safety issue, the investigator is advised to wear protective clothes from head to toe while sampling plants[106] for scientific analysis to prevent contamination or harm to oneself, either in the field or indoor environment. Moldy places contain potential hazards for investigators or crime scene personnel during short- or long-term investigations, as in cases such as homicide and sexual assault. The hazard results from breathing, searching, and moving and touching items covered with mold spores and dust. The examiner, whether working at the crime scene or in the lab, whose hands are unprotected (i.e., with gloves) may become a source of fungal infection from hairs, desquamated skin epithelium, clothes, or body contact with an infected person.[18] It is also important to take precautions when working with soil or with environmental materials that can harbor infectious conidia (spores) such as *Coccidioides immitis*.[73] Bioterrorism case investigations require additional precautions and protection measures.

Macroscopic plant evidence such as mushrooms or mold growth is easily recognized and of high value. However, in cases of microscopic evidence or in which the probative value of fungal growth is unclear, an investigator's awareness, education,[105] and community culture, as well as the crime scene circumstances, deterioration level, epidemiological or other investigation requirements will play a role in the sampling decision. Following routine investigation standards and sampling and analyzing procedures will produce data that will not always be critical or necessary, but it will reduce the possibility of omitting important evidence for the case in which fungal impact is relevant.[107]

In poisoning cases involving mushrooms, the number of mushrooms eaten and preparation methods before ingestion (e.g., boiled or fresh) are important. The time between ingestion and onset of symptoms is vital to know, because most deadly types produce an effect after six hours.[1] In some cases of poisonous mushrooms, samples may not be available, due either to complete consumption or the victim's state of consciousness, and the victim (if possible) or companions may be asked to describe, draw, look at photos, or bring some similar mushroom samples for identification.[35]

Searching the place for illicit or poisonous mushrooms is important to recover any piece of evidence that can lead investigators to determine the species of fungi. The investigator may examine cooking tools, garbage baskets, field collection containers, sinks, tables, floor, oven, refrigerator, living rooms, food dishes, and the remains of cooked meals. Cultivation pots discovered in the home or at the crime scene could be used to grow hallucinogenic fungi and should be investigated. The growth may be obvious or unseen in its early stages, in which case, incubation and growth of the mushroom are needed to determine the species.[33]

Generally, the investigator should pack or wrap the mushroom evidence in paper[108] after drying, taking care to avoid mixing the items together, but keeping them separate according to the collection place.[105] Sampling from native environments should include the whole plant (caps and stalks) and also soil from underneath the plant and control samples for heavy metal analysis.[29] Investigators may use plastic bags when collecting plant samples (fungi) from the field, and include in the bag a nonreactive, clean paper slip with written information for documentation.[106] Cooling the sample is necessary if the weather is hot or if the analysis will take time, to avoid spoilage factors that will damage the samples.[109]

It is necessary to be cautious when working with food mycotoxin investigation cases and to look at the complete food production process — preharvest (agriculture methods), harvesting operations (cleanup and drying), and postharvest procedures (storage method and conditions) — to reach an accurate conclusion to the main cause of the problem. A mistake could happen as a result of actions by the farmer, trader, or from the consumer himself. In addition, following national or international law regarding the detected mycotoxin levels should be considered.[110]

Filamentous fungi growth can be detected by different methods, such as examining the microbial volatile organic compounds (MVOCs) in indoor samples to locate the sampling area.[111] An electronic nose (sniffer) method, which is based on conductive polymers, has been used to detect the toxigenic strains of *Fusarium* spp.[112] An array biosensor has been developed to detect many types of mycotoxins and other components; this is a rapid way for food-borne contaminations to be detected.[113] A trained dog appears to be highly specific and has a positive predictive value in the detection of some types of molds.[114]

One of the novel methods of detection of a microorganism is the TIGER (Triangulation Identification for the Genetic Evaluation of Risks) method, which depends on intelligent PCR primers to target microbial genomes and high-performance mass spectrometry measurements to analyze PCR products. TIGER is used to detect microorganisms that are derived from clinical samples, air filtration devices, or other sources.[115] There are many commercial testing kits available on the market for detection of some mold species and mold toxins in the field or in the laboratory. An example of such a kit is the QuickTox Kit for aflatoxin detection at 10 parts per billion from EnviroLogix Inc (www.envirologix.com), which was approved by the USDA's Grain Inspection Packers and Stockman's Administration (GIPSA).

Sampling fungi can be done by various methods such as sterile swabs,[52] liquid vials, and adhesive strips, which are fast, inexpensive, reproducible, and nondestructive ways for sampling fungi from liquids or other surfaces.[116] A nondestructive procedure recommended for sensitive artifacts or evidence is using a sterile plastic scalpel to scrape the biofilm from the surface of the sampling area.[102] An agar plate can be used with different media for either outdoor[117] or indoor environments for sampling.[41] In the case of samples such as wood, concrete, paints, plaster mixtures, and paper, decayed cuts or pieces can be taken for analysis directly.[104] Vacuuming methods can be used also in sampling dust for fungus growth.[118] The Burkard seven-day Volumetric Spore Trap (Burkard Mfg. Co., Rickmansworth, UK) can be used to collect atmospheric spores.

Methods for identifying and studying fungi and their products are various, such as macroscopic observation, light or electronic microscopy, culturing, histology, toxicology, immunology, and DNA analysis, depending on the sample and the case requirement. Fungal spores and hyphae, and mushroom body parts are important taxonomical characteristics that will be considered in the first stage of identification by the mycologist. Molecular typing of fungi is an upcoming new field in forensic science and provides genus or species identification by sequencing and searching in a dataset for a DNA match.[111] To assist hospitals, where quick identification of a species in mushroom poisoning cases is critical, a computerized system has been suggested for listing the most important mushrooms and their descriptions, along with other vital information.[119]

A DNA approach is used now to differentiate between hallucinogenic and nonillicit species of the same genus, which is important when discussing plant evidence in legal proceedings.[120,121] It is worth considering that DNA methods can be successful with small quantities of mushroom samples.[104] Some methods have been developed to detect specific mycotoxin-producing fungi such as specific PCR primer pairs for OTA- producing fungi *Aspergillus ochraceus*,[122] trichothecene-producing species,[123] or opportunistic pathogenic fungi in clinical samples.[124] The Scientific Working Group on Microbial Genetics and Forensics guidelines can be followed when working with fungi for genetic identification and quality assurance.[125] A growing number of fungi have been completely sequenced, which will provide more opportunities for research in this field.[8]

Fungi mycotoxins and poisons can be detected toxicologically in samples such as blood, urine, brain, stomach contents,[29] and nonbiological substances by using testing techniques for screening and confirmation such as GC/MS, HPLC, TLC, and ELISA.[22] Psilocybin in hallucinogenic mushrooms also can be determined by reversed-phase liquid chromatography (HPLC) with fluorescence (FL) detection.[126] To ensure quality in the detection and analysis of mycotoxin as bioterrorism agents, investigators should follow the recommendations of the Scientific Working Group on Forensic Analysis of Chemical Terrorism.[127] Measurements of ergosterol (primary fungal membrane steroid) content in tested samples are used to quantify fungal biomass as a biomarker in various samples such as soils, building materials, house dust, and as a tool for studying the indoor fungi level related to health problems. Gas chromatography and mass spectrometry in the MS/MS mode also can be used for measurement of ergosterol levels.[104] Determining the growth rate by weighing the fungi mycelium is ineffective, because the fungi does not increase exponentially during the growth stage except in the early stages; this is in contrast with bacteria.[4]

In the forensic investigation of fungal infection of medical products, the epidemiological working plan for *Fusarium* keratitis outbreak of contact lenses can be used as an example. The recommended investigation methods included many steps and procedures, such as obtaining data by taking a patient's medical history and interviewing patients and ophthalmologists. *Fusarium* isolates also were taken from the patient environment and clinic, which were then genotyped by multilocus sequencing. In addition, contact lens solution bottles and manufacturing plants were sampled and analyzed to reach a clear position for the source of this infection.[19] Investigation of fungi and their chemical products used in bioterrorism will follow generally the current forensic microbiology investigation methods. The work consists of the forensic practices of evidence searching, detection, and handling, as well as intelligence information collection procedures, in addition to epidemiological research to identify the etiological agent of the attack.[71]

9.15 Conclusion

The identification and analysis of fungi can be a powerful tool in forensic science. Some of these tools are well known, such as forensic toxicology (used in hallucination and mushroom poisoning), but others appear to be new or emerging areas of study, including fungal genetic identification by DNA, deterioration, hydrocarbons, postmortem interval estimation, mycotoxin, and bioterrorism, palynology, infection epidemiology, and space research. Fungi forensics holds many promises and challenges in practice, which will require more research, standardization, and harmonization with forensic investigation concepts of transfer principles, identification, classification, individualization, association, and reconstruction.[128] In addition, fulfilling the court requirements for evidence admissibility (e.g., *Frye* and *Daubert v Merrell Dow* standards) will make fungi evidence more relevant in providing case linkages.[129,130] On the other hand, fungi and their products could have harmful effects on some forensic work, which needs to be considered as equally important in estimating evidence quality, PMI, and storage conditions.

Acknowledgments

The author gratefully thanks Dr. Thomas Parsons and Ms. Cheryl Katzmarzyk for permission to use the photographs in Figure 2 and scientific assistance, respectively (International Commission on Missing Persons [ICMP], Bosnia & Herzegovina). I am thankful to Dr. Joan W. Bennett (Rutgers University, USA) for her support and opinions. For the references collection assistance, I would like to thank the Libraries Deanship (United Arab Emirates University) and Ms. Karen Neves (Dalhousie University, Canada). I thank also Mr. Brien Holmes (Emirates Natural History Group) for reading and correcting the manuscript.

References

1. Berger, K.J. and Guss, D.A., Mycotoxins Revisited. Part I, *J. Emerg. Med.*, 28, 53–62, 2005.
2. Heckman, D.S. et al., Molecular evidence for the early colonization of land by fungi and plants, *Sci.*, 293, 1129–1133, 2001.
3. Lutzoni, F. et al., Assembling the fungal tree of life: progress, classification, and evolution of subcellular traits, *Am. J. Bot.*, 91, 1446–1480, 2004.
4. Dantigny, H., Guilmart, A., and Bensoussan, M., Basis of predictive mycology, *Int. J. Food Biol.Microbiol.*, 100, 187–196, 2005.
5. Maurois, A., *The Life of Sir Alexander Fleming, Discoverer of Penicillin*, Hopkins, G., Trans., E. P. Dutton & Co. Inc., New York, 1959, p. 131.
6. Doyle, A.C., The musgrave ritual, in *The Complete Sherlock Holmes*, Geddes & Grosset, New Lanark, Scotland, 2002, p. 168.
7. Coyle, H.M. et al., The Green Revolution: botanical contributions to forensics and drug enforcement, *Croat. Med. J.*, 42, 340–345, 2001.
8. Galagan, J.E. et al., Genomics of the fungal kingdom: insights into eukaryotic biology, *Genome Res.*, 15, 1620–1631, 2005.
9. Osiewacz, H.D., Genes, mitochondria and aging in filamentous fungi, *Ageing Res. Rev.*, 1, 425–442, 2002.
10. Szczepanowskaa, H. and Cavaliereb, A.R., Fungal deterioration of 18th and 19th century documents: a case study of the Tilghman Family Collection, Wye House, Easton, Maryland, *Int. Biodet. Biodeg.*, 46, 245–249, 2000.
11. Erzinçliolu, Z., *Blowflies,* Naturalist's Handbook 23, The Richmond Publishing Co. Ltd, Slough, UK, 1996, pp. 16–25.
12. Richardson, M.D., Changing patterns and trends in systematic fungal infections, *J. Antimicrob. Chemother.*, 56, 5–11, 2005.
13. Patterson, T.F., Advances and challenges in management of invasive Mycoses, *Lancet*, 366, 1013–1025, 2005.

14. Wetli, C.V. et al., Fungal cerebritis from intravenous drug abuse, *J. Forens. Sci.*, 29, 260–268, 1984.
15. Wilson, C.E., Sudden infant death syndrome and Canadian Aboriginals: bacteria and infections, *FEMS Immunol. Med. Microbiol.*, 25, 221–226, 1999.
16. Jarvis, B.B., *Stachybotrys chartarum*: a fungus for our time, *Phytochem.*, 64, 53–60, 2003.
17. Bennett, J.W. and Klich, M., Mycotoxins, *Biol. Microbiol. Clin. Microbiol. Rev.*, July, 497–516, 2003.
18. Kwon-Chung, K. J. and Bennett, J.E., *Medical Mycology*, Lea & Febiger, Malvern, Pa, 1992.
19. Douglas, C. et al., Multistate outbreak of *Fusarium* Keratitis associated with use of contact lens solution, *JAMA*, 296, 23–30, 2006.
20. Chen, J. et al., Characteristics of fungal growth in soft contact lenses, *ICLC*, 26, 84–91, 1999.
21. Serra, R., Braga, A., and Venâncio, A., Mycotoxin-producing and other fungi isolated from grapes for wine production, with particular emphasis on ochratoxin A, *Res. Microbiol.*, 156, 515–521, 2005.
22. Anklam, E., Stroka, J., and Boenke, A., Acceptance of analytical methods for implementation of EU legislation with a focus on mycotoxins, *Food Cont.*, 13, 173–183, 2002.
23. Puntarić, D. et al., Ochratoxin A in corn and wheat: geographical association with endemic nephropathy, *Croat. Med. J.*, 42, 175–180, 2001.
24. Speijers, G.J.A. and Speijers, M.H.M., Combined toxic effects of mycotoxins, *Toxicol. Lett.*, 153, 91–98, 2004.
25. Bennett, J.W., personal communication, 2005.
26. Domijan, A.-M. et al., Seed-borne fungi and ochratoxin: a contamination of dry beans (*Phaseolus vulgaris L.*) in the Republic of Croatia, *Food Chem. Toxicol.*, 43, 427–432, 2005.
27. WHO, Expert group meeting on Aflatoxins and health, Brazzaville, Congo, 24–27 May 2005, Regional Office for Africa, Brazzaville—Congo. www.afro.who.int/des/meetings/expert_grp_meet_may05_aflatoxins.pdf.
28. Koh, S. et al., Rapid detection of fungal endophytes in grasses for large-scale studies, *Funct. Eco.*, 20, 736–742, 2006.
29. Falandysz, J. et al., Mercury in wild mushrooms and underlying soil substrate from Koszalin north-central Poland, *Chemos.*, 54, 461–466, 2004.
30. Sadler, M., Nutritional properties of edible fungi, *Nutr. Bull.*, 28, 305–308, 2003.
31. Bonnet, M.S. and Basson, P.W., The toxicology of *Amanita phalloides*, *Homeop.*, 91, 249–254, 2002.
32. Alexopoulos, C.J. and Mims, C.W., *Introductory Mycology*, 3rd ed., John Wiley & Sons, Singapore, 1979.

33. Gonmori, K. and Yoshioka, N., The examination of mushroom poisonings at Akita University, *Legal Med.*, 5, S83–S86, 2003.
34. Alves, A. et al., Mushroom poisoning with *Amanita phalloides*, a report of four cases, *Eur. J. Intern. Med.*, 12, 64–66, 2001.
35. Özçay, F. et al., Fulminant liver failure secondary to mushroom poisoning in children: importance of early referral to a liver transplantation unit, *Pediat. Transplant.*, 10, 259–265, 2006.
36. Halpern, J.H., Hallucinogens and dissociative agents naturally growing in the United States, *Pharmacol. Ther.*, 102, 13–138, 2004.
37. Musshoff, F., Madea, B., and Beike, J., Hallucinogenic mushrooms on the German market simple instructions for examination and identification, *Forens. Sci. Int.*, 113, 389–395, 2000.
38. Tsujikawa, K. et al., Morphological and chemical analysis of magic mushrooms in Japan, *Forens. Sci. Int.*, 138, 85–90, 2003.
39. Satora, L. et al., Fly agaric Amanita muscaria poisoning, *Toxicon.*, 45, 941–943, 2005.
40. Immonen, J. et al., Skin-prick test findings in students from moisture-and mould-damaged schools: a 3-year follow-up study, *Pediatr. Allergy Immunol.*, 12, 87–94, 2001.
41. Portnoy, J.M., Barnes, C.S., and Kennedy, K., Sampling for indoor fungi, *J. Allergy Clin. Immunol.*, 189–198, 2004.
42. Shoemaker, R.C. and House, D.E., Sick building syndrome (SBS) and exposure to water-damaged buildings: time series study, clinical trial and mechanisms, *Neurotoxicol. Teratol.*, 28, 573–588, 2006.
43. Kildesø, J. et al., Determination of fungal spore release from wet building materials, *Indoor Air*, 13, 148–155, 2003.
44. Brasel, T.L. et al., Detection of airborne *Stachybotrys chartarum* macrocyclic trichothecene mycotoxins on particulates smaller than Conidia, *Appl. Environ. Biol.Microbiol.*, 71, 114–122, 2005.
45. Brandt, M. et al., Mold prevention strategies and possible health effects in the aftermath of hurricanes and major floods, *MMWR*, 55, 1–27, 2006.
46. United States Cong. 108th Congress, 1st Session, H.R. 1268, *United States Toxic Mold Safety and Protection Act of 2003 or The Melina Bill.* Introduced by Mr. Conyers, the House of Representatives, 13 March 2003.
47. Lehtomäki, K. et al., Risk analysis of Finnish peacekeeping in Kosovo, *Risk Anal.*, 25, 389–396, 2005.
48. Ishii, K. et al., Analysis of fungi detected in human cadavers, *Legal Med.*, 8, 188–190, 2006.
49. Carter, D.O. and Tibbett, M., Taphonomic Mycota, fungi with forensic potential, *J. Forens. Sci.*, 48(1), 168–171, 2003.
50. Bunyard, B.A., Commentary on: Carter, D.O., Tibbett, M., Taphonomic Mycota, fungi with forensic potential, *J. Forens. Sci.*, 48(1), 168–71, 2003.

51. Hitosugi, M. et al., Fungi can be a useful forensic tool, *Legal Med.*, 8, 240–242, 2006.
52. Hammad, E.F., Al-Janabi, A.A., and Mohamed, S.A., Fungi that grow on formalin-fixed cadavers, *Saudi Med. J.*, 23, 871–872, 2002.
53. Mairs, S., Swift, B., and Rutty, G.N., Detergent: an alternative approach to traditional bone cleaning methods for forensic practice, *Am. J. Forens. Med. Pathol.*, 25, 276–284, 2004.
54. Piepenbrink, H., Two examples of biogenous dead bone decomposition and their consequences for taphonomic interpretation, *J. Archaeol. Sci.*, 13, 417–430, 1986.
55. Trueman, C.N. and Martill, D.M., The long term survival of bone: the role of bioerosion, *Archaeom.*, 44, 371–382, 2002.
56. Marchiafava, V., Bonucci, E., and Ascenzi, A. Fungal osteoclasia: a model of dead bone resorption, *Calcif. Tiss. Int.*, 14, 195–210, 1974.
57. Katiyar, S. and Kushwaha, R.K.S., Invasion and biodegradation of hair by house dust fungi, *Int. Biodet. Biodeg.*, 50, 89–93, 2002.
58. DeGaetano, D.H., Kempton, J.B., and Rowe, W.F., Fungal tunneling of hair from a buried body, *J. Forens. Sci.*, 37, 1048–1054, 1992.
59. Deacon, L.J. et al., Diversity and function of decomposer fungi from a grassland soil, *Soil Biol. Biochem.*, 38(1), 7–20, 2006.
60. Powell, K.L. et al., Ultrastructural observations of microbial succession and decay of wood buried at a Bronze Age archaeological site. *Int. Biodet. Biodeg.*, 47, 165–173, 2001.
61. Martín, A. et al., Contribution of a selected fungal population to the volatile compounds on dry-cured ham, *Int. J. Food Microbiol.*, 110, 8–18, 2006.
62. Martín. A. et al., Contribution of a selected fungal population to proteolysis on dry-cured ham, *Int. J. Food Microbiol.*, 94, 55–66, 2004.
63. Klintworth, R. et al., Biological induced corrosion of materials II: new test methods and experience from MIR station, *Acta Astronaut.*, 44, 569–578, 1999.
64. Gu, J.-D. et al., The role of microbial biofilms in deterioration of space station candidate materials, *Int.Int. Biodet. Biodeg.*, 41, 25–33, 1998.
65. Kirkbride, K.P. et al., Microbial degradation of petroleum hydrocarbons: implications for arson residue analysis, *J. Forens. Sci.*, 37, 1585–1599, 1992.
66. Okerentugba, P.O. and Ezeronye, O.U., Petroleum degrading potentials of single and mixed microbial cultures isolated from rivers and refinery effluent in Nigeria, *Afr. J. Biotechnol.*, 2, 288–292, 2003.
67. Ladygina, N., Dedyukhina, E.G., and Vainshtein, M.B., A review on microbial synthesis of hydrocarbons, *Process Biochem.*, 41, 1001–1014, 2006.
68. Lewis, R.J. et al., Ethanol formation in unadulterated postmortem tissues, *Forens. Sci. Int.*, 146, 17–24, 2004.

69. Yajima, D. et al., Ethanol production by *Candida albicans* in postmortem human blood samples: effects of blood glucose level and dilution, *Forens. Sci. Int.*, 164, 116–121, 2006.
70. Jelen´, H.H., Use of solid phase microextraction (SPME) for profiling fungal volatile metabolites, *Lett. Appl. Biol. Microbiol.*, 36, 263–267, 2003.
71. Budowle, B., Murch, R., and Chakraborty, R., Microbial Forensics: the next forensic challenge, *Int. J. Leg. Med.*, 119, 317–330, 2005.
72. CDC, Chemical Emergencies, case definition, *Trichothecene Mycotoxins*, 17, 1–2, 2005.
73. Dixon, D.M., *Coccidioides immitis* as a select agent of bioterrorism, *J. Appl. Biol. Microbiol.*, 91, 602–605, 2001.
74. Foxell, J.W., Trends in Bio-Terrorism: two generations of potential weapons, *J. Conting. Crisis Manag.*, 7, 102–118, 1999.
75. Butcher, J., Virulence factor identified for *Histoplasma capsulatum, Lancet*, 356, 1741, 2000.
76. Stahl, C.J., Green, C.C., and Farnum, J.B., The incident at Tuol Chrey: pathologic and toxicological examination of casualty after chemical attack, *J. Forens. Sci.*, 30, 317–337, 1985.
77. Bailey, B.A. et al., Evaluation of infection processes and resulting disease caused by *Dendryphion penicillatum* and *Pleospora papaveracea* on *Papaver somniferum*, *Phytopathol.*, 90, 699–709, 2000.
78. Poppy wars, news & comments, *Trends Plant Sci.*, 7, 338, 2002.
79. Mildenhall, D.C. et al., Forensic palynology: why do it and how it works (editorial), *Forens. Sci. Int.*, 163, 163–172, 2006.
80. Mildenhall, D.C., An unusual appearance of a common pollen type indicates the scene of the crime, *Forens. Sci. Int.*, 163, 236–240, 2006.
81. Benyon, F.H.L. et al., Differentiation of allergenic fungal spores by image analysis, with application to aerobiological counts, *Aerobiol.*, 15, 211–223, 1999.
82. Czeczuga, B. et al., Aquatic fungi growing on dead fragments of submerged plants, *Limnol.*, 35, 283–297, 2005.
83. Czeczuga, B., Godlewska, A., and Kozlowska, M., Zoosporic fungi growing on the carapaces of dead zooplankton organisms, *Limnol.*, 30, 37–43, 2000.
84. Garner, T.W.J. et al., The emerging amphibian pathogen *Batrachochytrium dendrobatidis* globally infects introduced populations of the North American bullfrog, *Rana catesbeiana. Biol. Lett.*, 2, 455–459, 2006.
85. Johnson, M.L. and Speare, R., Survival of *Batrachochytrium dendrobatidis* in Water: quarantine and disease control implications, *Emerg. Infect. Dis.*, 9, (8), 922–925, 2003.
86. Prasannarai, K. and Sridhar, K.R., Diversity and abundance of higher marine fungi on woody substrates along the west coast of India, *Curr. Sci.*, 81, 304–311, 2001.

87. Dykstra, M.J. and Kane, A.S., *Pfiesteria piscicida* and ulcerative mycosis of atlantic menhaden: current status of understanding, *J. Aquat. Anim. Health,* 12, 18–25, 2000.
88. Sallenave-Namont, C. et al., Toxigenic saprophytic fungi in marine shellfish farming areas, *Mycopathol.,* 149, 21–25, 2000.
89. Grovel, O., Pouchus, Y.F., and Verbist, J.F., Accumulation of gliotoxin, a cytotoxic mycotoxin from *Aspergillus fumigatus*, in blue mussel (*Mytilus edulis*), *Toxicon.*, 42, 297–300, 2003.
90. Burford, E.P, Fomina, M., and Gadd, G.M., Fungal involvement in bioweathering and biotransformation of rocks and minerals, *Mineral. Mag.*, 67, 1127–1155, 2003.
91. Szostak-Kotowa, J., Biodeterioration of textiles, *Int. Biodet. Biodeg.*, 53, 165–170, 2004.
92. Seves, A. et al., The microbial degradation of silk a laboratory investigation, *Int. Biodet. Biodeg.*,42, 192–100, 1998.
93. Wesenberg, D., Kyriakides, I., and Agathos, S.N., White-rot fungi and their enzymes for the treatment of industrial dye effluents, *Biotechnol. Adv.*, 22, 161–187, 2003.
94. Harazono, K. and Nakamura, K., Decolorization of mixtures of different reactive textile dyes by the white-rot basidiomycete *Phanerochaete sordida* and inhibitory effect of polyvinyl alcohol, *Chemosphere,* 59, 63–68, 2005.
95. Mohorčič, M. et al., Fungal and enzymatic decolourisation of artificial textile dye baths, *Chemos.*, 63, 1709–1717, 2006.
96. Orlita, A., Microbial biodeterioration of leather and its control: a review, *Int. Biodet. Biodeg.*, 53, 157–163, 2004.
97. Janaway, R.C., Degradation of clothing and other dress materials associated with buried bodies of archaeological and forensic interest, in *Advances in Forensic Taphonomy, Methods, Theory, and Archaeological Perspectives*, Haglund, W.D. and Sorg, M.H., Eds., CRC Press, Boca Raton, 2002, pp. 379–402.
98. Darby, R.T., Simmons, E.G., and Wiley, B.J., A survey of fungi in military aircraft fuel supply systems, *Int.Int. Biodet. Biodeg.*, 48, 159–161, 2001.
99. Bentoa, F.M. and Gaylarde, C.C., Biodeterioration of stored diesel oil: studies in Brazil, *Int.Int. Biodet. Biodeg.*, 47, 107–112, 2001.
100. Abrusci, C. et al., Isolation and identification of bacteria and fungi from cinematographic films, *Int.Int. Biodet. Biodeg.*, 56, 58–68, 2005.
101. Wasserbauer, R., Microbial biodeterioration of electrotechnical insulation materials, *Int. Biodet. Biodeg.*, 53, 171–176, 2004.
102. Videla, H.A., Guiamet, P.S., and De Saravia, S.G., Biodeterioration of Mayan archaeological sites in the Yucatan Peninsula, Mexico, *Int. Biodet. Biodeg.*, 46, 335–341, 2000.

103. Jellison, J. and Jasalavich, C., A review of selected methods for the detection of degradative fungi, *Int.Int. Biodet. Biodeg.*, 46, 241–244, 2000.
104. Hippelein, M. and Rugamer, M., Ergosterol as an indicator of mould growth on building materials, *Int. J. Hyg. Environ. Health,* 207, 379–385, 2004.
105. Coyle, H.M. et al., Forensic botany: using plant evidence to aid in forensic death investigation, *Croat. Med. J.*, 46, 606–612, 2005.
106. Cseke, L.J. et al., *Handbook of Molecular and Cellular Methods in Biol. and Medicine,* 2nd ed., CRC Press, Boca Raton, 2004, p. 511.
107. Sorg, M.H. and Haglund, W.D., Advancing forensic taphonomy: purpose, theory, and process, in Haglund, W.D. and Sorg, M.H., Eds., *Advances in Forensic Taphonomy, Methods, Theory, and Archaeological Perspectives,* CRC Press, Boca Raton, 2002, p. 19.
108. Lee, H.C., Palmbach, T.M., and Miller, M.T., *Henry Lee's Crime Scene Handbook,* Academic Press, San Diego, 2001, p. 177.
109. Lee, H.C. and Ladd, C., Preservation and collection of biological evidence, *Croat. Med. J.,* 42, 225–228, 2001.
110. Lopez-Garcia, R., Park, D.L., and Phillips, T.D., Integrated mycotoxin management systems, *FAO, Food, Nutr. Agric.,* 23, 38–48, 1999.
111. Rozynek, P. et al., Quality test of the MicroSeq D2 LSU fungal sequencing kit for the identification of fungi, *Int. J. Hyg. Environ. Health,* 206, 297–299, 2003.
112. Falasconi, M. et al., Detection of toxigenic strains of *Fusarium verticillioides* in corn by electronic olfactory system, *Sens. Actuat. B,* 108, 250–257, 2005.
113. Sapsford, K.E. et al., Rapid detection of foodborne contaminants using an array biosensor, *Sens. Actuat. B: Chem.,* 113, 599–607, 2006.
114. Kauhanen, E. et al., Validity of detection of microbial growth in buildings by trained dogs, *Environ. Int.,* 28, 153–157, 2002.
115. Hofstadler, S.A. et al., TIGER: the universal biosensor, *Int. J. Mass Spectrom.,* 242, 23–41, 2005.
116. Urzi, C. and De Leo, F., Sampling with adhesive tape strips: an easy and rapid method to monitor microbial colonization on monument surfaces, *J. Microbiol. Meth.,* 44, 1–11, 2001.
117. Fang, Z. et al., Culturable airborne fungi in outdoor environments in Beijing, China, *Sci. Tot. Environ.,* 350, 47–58, 2005.
118. Macher, J.M., Review of methods to collect settled dust and isolate culturable microorganisms, *Indoor Air,* 11, 99–110, 2001.
119. Zotti, M. et al., A decision support system for the management of accidental mushroom and plant poisoning, *Il Farmaco,* 56, 391–395, 2001.
120. Nugent, K.G. and Saville, B.J., Forensic analysis of hallucinogenic fungi: a DNA–based approach, *Forens. Sci. Int.*, 140, 147–157, 2004.

121. Lee, J.D, Cole, M., and Linacre, A., Identification of members of the genera *Panaeolus* and *Psilocybe* by a DNA test, a preliminary test for hallucinogenic fungi, *Forens. Sci. Int.,* 112, 123–133, 2000.

122. Dao, H.P., Mathieu, F., and Lebrihi, A., Two primer pairs to detect OTA producers by PCR method, *Int. J. Food Microbiol.,* 104, 61–67, 2005.

123. Jurado, M. et al., PCR detection assays for the trichothecene-producing species *Fusarium graminearum, Fusarium culmorum, Fusarium poae, Fusarium equiseti* and *Fusarium sporotrichioides, Syst. Appl. Biol. Microbiol.*, 28, 562–568, 2005.

124. Brancart, F. et al., Quantitative TaqMan PCR for detection of *Pneumocystis jiroveci, J. Microbiol. Meth.*, 61, 381–387, 2005.

125. SWGMGF, Quality assurance guidelines for laboratory performing microbial forensic work, *Forens. Sci. Comm.,* 5(4), 2003.

126. Saito, K. et al., Determination of psilocybin in hallucinogenic mushrooms by reversed-phase liquid chromatography with fluorescence detection, *Talanta*, 66, 562–568, 2005.

127. SWGFACT, Validation guidelines for laboratories performing forensic analysis of chemical terrorism, *Forens. Sci. Comm.,* 7, 1–14, 2005.

128. Inman, K. and Rudin, N., The origin of evidence, *Forens. Sci. Int.,* 126, 11–16, 2002.

129. Keierleber, J.A. and Bohan, T., Ten years after Daubert: the status of the states, *J. Forens. Sci.*, 50, 1–10, 2005.

130. Robertson, B. and Vignaux, G.A., *Interpreting Evidence; Evaluating Forensic Science in the Courtroom,* John Wiley & Sons, West Sussex, 1999.

10 Soil DNA Typing in Forensic Science

JONATHAN HILL, LINDA STRAUSBAUGH, AND JOERG GRAF[1]

Contents

10.1 Introduction

Forensic scientists are, first and foremost, in the business of analyzing evidence. From traditional fingerprinting to DNA-based genotyping, forensic techniques seek to physically link the various components of a crime together, such as tying the suspect to the crime scene or to the victim. Forensic science often borrows methodology from other sciences, such as biology and chemistry, and adapts biological and chemical methods to its own applications to solve crimes. Among the most successful innovations for the forensic science community was the use of modern genetic methods to compare evidentiary human DNA left at the scene of a crime. Genotyping techniques offer the advantages of speed, sensitivity, and compatibility with high throughput instrumentation. Additionally, statistical confidence factors and measurement

[1] The project was supported in part by Grant No. 2003-LP-CX-KO26, awarded by the National Institute of Justice, Office of Justice Programs, U.S. Department of Justice. Points of view in this chapter are those of the authors and do not represent the official position or policies of the U.S. Department of Justice.

of random match probabilities, made possible by large DNA databases, have further increased the forensic value of DNA.

As is apparent in this volume, there have been recent movements to extend genetic identification technology beyond human DNA to other forms of biological evidence. One sort of evidence especially suited to this approach is the analysis of soil samples. Soil contains a diverse community of microbes that can vary significantly in composition from one site to the next. Traces of soil evidence may be associated with suspects — under fingernails, on boots or clothing, or tracked into cars or homes. The analysis of trace soil from the suspect to link that individual to the scene of a crime establishes an important fact, one that could be of value for determining guilt or innocence. Some scientists believe that such forensic approaches could be of special use in wildlife crimes, such as poaching, where suspects almost always have direct contact with soil.[1] Horswell et al.[2] have provided proof of the concept that bacterial community DNA profiling can provide a new tool for the forensic analysis of soil samples. Using soil samples from footwear and clothing, as well as the impressions left at the sites, they were able to generate soil microbial community profiles by a standard T-RFLP profiling method. The profiles they obtained were representative of the site of collection and demonstrated the potential for use as associative evidence. This chapter explores the use of soil comparisons in forensic science, with an emphasis on modern genomic techniques that provide profiles of microbial communities. Methodological approaches are compared to provide an overview for forensic applications; foundational research and future directions are considered as well.

10.2 Historical Methods of Soil Analysis

The use of soil evidence certainly did not begin with modern DNA analysis techniques. Historically, the study of the physical characteristics of soil (e.g., mineral composition, size, color, shape) has been a discipline more closely tied to geology than genetics. These techniques were often limited by requirements of large sample sizes or by specialized expertise not resident in all forensic science laboratories. It can be argued that, although some physical analysis techniques are difficult, there are basic methods that require little equipment or expertise, and can be performed by any forensic laboratory on small samples.[3]

Foremost among such tests is the analysis of soil color, which can differ dramatically from site to site. Organic matter has a tendency to darken soils, while iron oxides can lend color ranging from yellow to brown or red. Measurement of soil color is performed by comparing the sample with a standard color chart. The soil sample can be small, significantly less than a gram, and as little as 50 milligrams for some tests. Soil may be simply air-dried and compared with the chart, or pretreated prior to comparison through moistening, organic matter decomposition, clay fractioning, iron oxide removal, or ashing. A single color measurement of an air-dried sample differentiated 70% of soil samples, and has a significant advantage in that some of the color tests, such as simple analysis of air-dried or moistened soil, do not consume the soil sample.[3]

Another relatively simple analytical test is sieving, that results in the separation of soil particles and a subsequent comparison of particle size profiles between the two samples. Sieving can be performed on amounts as small as 1 gram of soil, and requires little in the way of equipment. Data from sieving experiments have been found to discriminate between 95.9% of compared soil samples. When combined with color analysis, physical techniques can often draw distinctions between 99.5% of soil samples, which is impressive given the relatively simple nature of these two approaches.[4]

Although such analyses have the advantage of requiring simple equipment and training, recent improvements to other physical methods provide even greater discrimination. Among these is the recent use of Fourier transform infrared spectroscopy (FTIR) to analyze soil samples and improve upon color distinctions. In this method, an IR spectrum is taken of a soil sample; following degradation of the organic component by oxidative pyrolysis, a second spectrum is taken. By comparing these two spectra, the scientist can reconstruct the spectrum of the organic portion of the soil, which contains the information most useful for drawing distinctions between different soils. FTIR analysis is more sensitive to differences between soils than simple color comparison, but also requires specialized equipment and expertise. Still, it demonstrates that physical comparisons between soils are not necessarily outdated and uninformative compared to molecular techniques,[1] quite the contrary.

Basic microbiological culture methods were also used to compare soils before molecular biology techniques became commonplace. Typically, isolates (bacteria and fungi) were collected and grown on agar plates to compare colony morphology between soil samples.[5] Approximately 1 gram of soil would be suspended in buffer, diluted, and spread onto sets of agar plates containing different nutrients. This approach was sound, but required microbial expertise unlikely to be resident in every forensic science laboratory. With this approach, it was often difficult to quantify differences and one must rule out contamination as a potentially important source of error. Nevertheless, studies like these provided a foundation for forensic scientists to begin the consideration of microbes as sources of valuable forensic information, a view necessary to the later development of molecular methods.

10.3 Microbial DNA Typing Strategies

DNA-based methods of soil comparison are attracting the most recent attention from forensic specialists. DNA has taken the forensic science world by storm, and many laboratories have moved into the area on a grand scale, employing automated sequencers that provide sensitivity, scale, and reproducibility unprecedented in identification science. Human DNA typing techniques are based upon using a set of approved and validated short tandem repeat markers (STRs) that provide a probability of a random match in a population that is extremely small for anyone other than identical twins. If soil analysis could be complemented by DNA methods, the technology would be easy to adopt and use, given that forensic scientists are both comfortable with DNA technology and highly competent in its use.

Microbial ecologists have been developing DNA techniques for typing soil microbial communities for several years. Frustrated by the difficulty of culturing many soil microbes in the laboratory, they have sought to establish general methods by which all soil microbes could be analyzed. Broadly speaking, such techniques involve extracting the DNA present in a small sample of soil and, in all but a few cases, amplifying a segment of DNA by the polymerase chain reaction (PCR) for further analysis.

10.4 DNA Extraction from Soil

Extraction of DNA from soil samples is seemingly straightforward, but is actually one of the most critical and problem-ridden steps microbial DNA typing protocols. The major problem is that DNA extraction from soil often includes contaminants such as humic acid and other PCR inhibitors.[6] Additionally, the DNA extraction efficiency varies significantly between different types of soil, such as forest soil and wetland sediments.[7] There is no ideal, universally applicable, method of DNA extraction for all soil types, as they differ significantly in composition.[8] Generally, direct DNA extraction from soil samples is preferable to the isolation of whole cells, followed by cell lysis and subsequent DNA extraction.[8] Although direct DNA extraction produces a better yield of DNA in a shorter time, it also tends to extract more chemical inhibitors, which must then be removed prior to DNA amplification.[8]

All soil DNA extraction protocols employ a lysis step to disrupt cells. Chemical lysis typically combines detergent and lytic enzymes to rupture the cells and release the DNA. Mechanical lysis is more effective and less selective although it risks shearing the DNA into smaller pieces. For most applications, moderate shearing is not overly detrimental to achieving a DNA profile. Mechanical lysis typically involves bead milling, freeze-thawing, or ultrasonication[8] as a method to break cells open. One set of experiments to evaluate methods for extraction of DNA from soils determined that bead mill homogenization in a lysis mixture (chloroform, SDS, NaCl, and phosphate-Tris, pH 8) is the best physical method in terms of DNA yield and cell lysis efficiency.[7] Freeze-thawing of the soil sample was found to be significantly less effective.

Many soil extraction protocols produce crude DNA samples that cannot be analyzed by PCR without further purification. One comparative study found a Sepharose G-200 spin column purification superior to other standard methods (spin-bind column purification, gel electrophoresis, and ammonium acid precipitation) for removing inhibitors while retaining DNA.[7] The following additional steps may be taken to further improve DNA amplification:

1. Additional purification may be performed with hexadecyltrimethylammonium bromide (CTAB) and polyvinylpolypyrrolidone.[7]
2. Samples may be diluted before PCR to lessen the impact of inhibitors.
3. The common practice of adding sequestering agents — such as bovine serum albumin (BSA) or other proteins — may be used to lessen the effect of inhibitors.[8]

Soil extraction has been made far easier with the availability of commercial kits that offer reagents in a convenient quality-controlled, disposable format. MoBio Laboratories Inc. supplies one popular kit, the Ultraclean Soil DNA Isolation kit, designed to extract DNA from 0.25 to 1.00 gram of soil, using a simple protocol. Epicentre manufactures a competing kit, the Soil-Master kit, which involves a hot-detergent lysis step. Kit-based methods are highly desirable for forensic applications. Although standardized approaches may preclude the development of other optimal methodology for different soil types, having standard protocols easily exchangeable between laboratories in an easy-to-use format is desirable.

10.5 Methods for Typing Soil-Based DNA Samples

One of the earliest techniques used to compare soil sample DNA was G + C typing.[9] Microbiologists find the quantity of guanine plus cytosine in extracted DNA useful for coarse-level typing of soils, when a rough comparison is desired. This metric can sometimes provide taxonomic information on the organisms present in the soil as well. One advantage of G + C typing is that it is one of the few techniques available for soil-DNA comparison that does not employ PCR at any stage, and is therefore neither subject to biases that amplification steps might introduce nor affected by PCR inhibitors.

However, G + C typing has a number of drawbacks that render it unsuitable for forensic applications. The method is lengthy (about four days) and requires an ultracentrifuge. More importantly, G + C typing requires a relatively large quantity of DNA, on the order of 50 micrograms. Finally, it cannot distinguish between different samples with the sensitivity of PCR-based methods, a potentially critical limitation for forensic purposes.

Denaturing gradient gel electrophoresis (DGGE) and temperature gradient gel electrophoresis (TGGE) are related techniques that are highly sensitive and find wide use among microbial ecologists.[9] Both techniques amplify the genes encoding the 16S rRNA with long primers (60-mers) designed to contain regions of high G + C content that form a stabilizing "clamp" to keep the DNA from melting easily. The PCR products are then analyzed on specialized gel systems that resolve DNA fragments of differing sequences based on their differential melting behaviors.[10] In DGGE, a gradient gel made with denaturants, such as urea or formamide, is used.[8] TGGE uses a linear temperature gradient.

Microbiologists using DGGE occasionally prefer it to other methods because fragments representing different members of a microbial community can be identified through DNA sequencing of excised bands or through hybridization with select probes.[10] Another advantage of the technique is that it can be adapted to use fluorescent dyes, which provide advantages in greater sensitivity and have the potential to be multiplexed. DGGE/TGGE methods have a number of disadvantages that are detrimental to adoption for forensic studies. First, the specialized gel systems are neither resident in most forensic science laboratories, nor easily adaptable to automated sequencing instruments. The long primers can cause artifacts and result in primer-dimer formation,[11–13] which can complicate the interpretation of the results. Most significantly, the lack of appropriate size standards for DGGE and TGGE makes comparison between gels difficult,[9,13] and would frustrate efforts to create a forensic database built on these techniques.

Single-strand conformation polymorphism (SSCP) is a method widely used by scientists studying both microbial communities and mutations. For microbial community comparisons, the technique depends upon PCR amplification of DNA extracted from soil. Most typically, primers are chosen to amplify the 16S rRNA gene, and amplicons are run on a gel in single-stranded forms that assume folded structures determined by intramolecular interactions under nondenaturing conditions. Mobility through the gel is determined by length, molecular weight, and shape. SSCP was sensitive enough to detect a population comprising fewer than 1.5% of a bacterial community when it was first adapted to microbial community typing.[11] Products may be visualized by silver-staining[12] or by fluorescence-detecting automated sequencers.[14] The original SSCP experiments produced three bands for each PCR product: two bands representing single-stranded DNA and one representing double-stranded DNA. Later modifications were developed to result in a single strand, eliminating this complication to result interpretations (12). There are a number of limitations to SSCP. There is no way to accurately define the melting temperatures of the helices, resulting in problems of quantifying data for inclusion into a database.[9] Some single-stranded nucleic acids can exist in several stable conformations and produce different bands, confounding output without providing useful information.[9] Variation in ambient temperature can also affect the results of SSCP runs, so it is important that the equipment be carefully temperature controlled[14] within and between laboratories.

Random amplified polymorphic DNA (RAPD) analysis is another PCR-based technique for comparing DNA extracted from soil but, unlike the previous approaches, it does not rely on primers that amplify the 16S rRNA gene. Instead, amplification is accomplished with short primers of arbitrary sequence that amplify random portions of the template DNA. Such short primers typically anneal at many sites throughout the genome and will generate a series of 5 to 15 distinct bands on an agarose gel. Frequently, the technique is repeated with 10 to 15 different primers so that statistical comparison between samples is more meaningful,[15] but smaller numbers of primers are sometimes used.[16] RAPD technology can be combined with fluorescent labeling and fragment analysis on automated DNA sequencers for high throughput analysis of samples.[17] Improvements in band resolution mean that fewer primers can be used to achieve comparable statistical relevance.

RAPD technology has the advantage of avoiding 16S primers, which may not bind with equal efficiency to all species of a microbial community in a soil sample.[15] Exclusive use of 16S rDNA-based techniques can lead scientists to a somewhat skewed picture of the diversity within the community. However, this issue is more relevant for those directly studying microbial ecology than for forensic scientists. Forensic applications are most often strictly concerned with comparing soil samples, and a 16S primer bias is not important to the interpretation of results, as long as that bias is applied uniformly to all samples typed by the same method. The requirement of multiple reactions with different primers can be an important consideration for forensic laboratories, since each PCR reaction requires use of additional template DNA, which may be a scare commodity.[15] In addition, there are concerns about the reproducibility of RAPD results between laboratories, a serious issue in the validation of the method.

Amplified fragment length polymorphism, AFLP, offers many of the advantages of RAPD analysis, with few of its drawbacks. Compared to other techniques for soil DNA comparison, the methodology of AFLP is complex, but the advent of commercial kits makes it simple in practice. To perform AFLP, DNA extracted from soils is digested with two restriction endonucleases, such as *Eco*RI and *Mse*I. Fragments of DNA that serve as adaptors are then ligated to the ends of the DNA fragments to generate the template DNA for subsequent PCR steps. The protocol is simplified considerably by the fact that the digestion and ligation steps can occur in a single reaction mixture. Preselective amplification is used to amplify the genomic fragments that have adaptors on both ends. At this stage, an arbitrary nucleotide may be added to the end of the primers to reduce the number of amplified sequences and simplify the final DNA pattern. Finally, a second round of PCR known as selective amplification is performed with fluorescent-dye-labeled primers for detection of the DNA fragments. AFLP adapts easily to most automated DNA sequencing instruments and software. However, the standard kits for performing AFLP profiling on bacteria are intended for analysis of one bacterial type at a time. If run on a community sample, there are often too many bands to easily make comparisons between samples. Some approaches avoid this problem by using primers designed for larger genomes (such as those of plants), which bind less frequently and provide a less complex banding pattern for microbial communities.[18] Using PCR primers to vary the sensitivity of the technique can make it applicable to typing large microbial communities, as well as specific strains of bacteria. Like RAPD analysis, AFLP avoids the use of PCR primers that target the 16S rRNA and presents a less-biased view of the microbial community within the soil sample.[19]

Amplified rDNA restriction analysis (ARDRA) is an approach to sample analysis that finds little use in modern labs, but was the precursor to T-RFLP analysis. To perform an ARDRA analysis, the DNA extracted from soil is amplified using primers for the 16S rRNA gene. Then, the amplification products are digested with a restriction enzyme and displayed on a gel.[9] ARDRA can be used to demonstrate the presence of specific phylogenetic groups or for estimating the overall species richness and distribution within a sample[20] The disadvantage of ARDRA is that results can be complex and difficult to interpret, especially if the soil has a diverse microbial community.[9] In addition, since the technique relies on the digestion of PCR products, ARDRA in the strict sense cannot be adapted to automated DNA sequencers, since not all DNA fragments can be fluorescently labeled.

Terminal restriction fragment length polymorphism or T-RFLP is an adaptation of the ARDRA technique.[20] Due to its relative ease of use and high sensitivity, T-RFLP finds the widest use of any microbial typing technique among scientists in the field today. Recently, Horswell et al.[21] have proposed that bacterial community DNA profiling may provide a new tool for the forensic analysis of soil samples. Using soil samples from footwear and clothing, as well as the footwear impressions left at the sites, they were able to generate soil microbial community profiles by a standard T-RFLP profiling method.[20] The profiles were representative of the site of collection and demonstrated the potential for use as associative evidence.[21]

Like ARDRA, extracted DNA is amplified with primers (most commonly for the 16S gene) and the PCR amplification products are digested with restriction enzymes (usually with a 4 base pair recognition sequence). However, in T-RFLP, only the fluorescently labeled terminal fragments are detected.[20] The typical range of fragments that can be accurately analyzed varies from 50 to 600 bases.[9] Detection of only the terminal fragments makes the interpretation of the results significantly easier, and the incorporation of fluorescence makes T-RFLP highly sensitive.[20] The T-RFLP approach, because of its popularity, has undergone intense scrutiny by the microbiological community, a distinct advantage for adoption to casework and eventual court admissibility. Original studies found that the 5′ terminal fragment, generated by restriction enzyme digestion with *Hha*I or *Msp*I, yielded the most useful taxonomic information.[20] Later modifications[22] improved the accuracy of microbe community characterization by expanding the number of enzymes used (for example, *Sau*3A, *Hae*III, *Msp*I, and *Hha*I). Another study suggested that the resolving power of a particular restriction enzyme varies as the richness of a microbial community changes in the sample.[23] T-RFLP was deemed highly reproducible and robust once standardized, but results were sensitive to the use of different *Taq* polymerases or different annealing temperatures in the PCR step.[24] Some "pseudo-T-RF" artifacts, bands due to missed terminal restriction sites in incomplete digestions, were noted with some of the restriction enzymes; modifications can be incorporated to reduce this effect.[25]

Generally, T-RFLP analysis has been found to be rapid and sensitive in practice.[26] It has been applied to a variety of samples, including industrially polluted areas, parks, and brooks,[27] human blood,[28] deer fecal pellets and sand,[22] and the gut DNA of a termite.[20] Relatively pure samples of pathogens in blood were identifiable to the genus or species level, and T-RFLP adapted well to high-throughput needs, with a turnaround time of about 8 hours for sample processing.[28] The extension of T-RFLP analysis to capillary-based DNA sequencers has further improved throughput and sensitivity.[29] The most interesting development for the future of T-RFLP analysis has been the extension to genes other than the 16S rRNA genes, such as mercury resistance (*mer*) genes.[27] Using T-RFLP to identify microbial communities, it was shown that the *mer* gene had greater presence in industrially contaminated sites.[27] T-RFLP analysis can be applied to additional genes previously analyzed by normal restriction fragment length polymorphism (RFLP), such as genes for denitrification.[30] Using functional genes in addition to the standard 16S T-RFLP will provide improved power of discrimination, a consideration especially relevant to forensic applications.

One of the main barriers to the application of T-RFLP analysis in forensic science is the lack of standardization. The PCR cycling conditions used for T-RFLP vary from study to study[19,20,22,24,27] and these differences could have significant effects if comparisons were to be drawn from T-RFLP analysis between laboratories.[24] The battery of restriction enzymes required for analysis would need to be standardized as well, and the forensic community would need to decide how many digestions would be required to achieve a match between samples with enough weight to be useful in court, without unduly consuming evidence. Unlike AFLP, kits for T-RFLP are not yet available and most current laboratory protocols involve a mixture of PCR kits and outside reagents, such as the specially labeled primers.

One final method of analysis of potential use to forensic scientists is rRNA intergenic spacer analysis (RISA). Unlike many protocols that use PCR primers to amplify the gene for the small ribosomal subunit, RISA analysis amplifies the intergenic region located between the genes for the small and large ribosomal subunits. This region, known as the IGS, is highly variable in size between bacterial groups, ranging from 50 base pairs to more than 1.5 kb.[31] The amplification products are run on a gel and silver-stained.[32] An adaptation of the technique, automated RISA (A-RISA), uses fluorescently labeled primers for the PCR amplification and conducts the final analysis on an automated DNA sequencer. A-RISA is both highly sensitive and reproducible.[33] A microbial community may produce hundreds of distinct bands with the RISA technique.[32] The protocol is very simple, and unlike other methods with similar sensitivity, there is no need for enzymatic digestion steps. Since it employs PCR primers targeted to a specific region of DNA that is not absolutely conserved across all bacteria, it is subject to biases similar to those involved with methods that rely on PCR amplification of the 16S rRNA gene itself. In some cases, researchers have used this fact to advantage by targeting sequences found only in the bacteria of interest for the study.[34] The result is a simplified RISA gel that is easier for the researcher to interpret.

10.6 Methods Suitable to the Forensic Analysis of Soil Evidence

Research in soil community ecology has provided a variety of DNA-based techniques for forensic scientists to consider. In general, the criteria for success in both fields are the same: good methods must be sensitive and reproducible. Forensic validation places additional demands on reproducibility between different laboratories, especially if databases are to be constructed. It also requires that techniques have quick turnaround times and can be adapted to existing high-throughput sample needs. Techniques used by forensic scientists must be capable of functioning on very small sample sizes. However, unlike methods applied to microbial ecology, the techniques chosen by forensic scientists do not require the absolute absence of artifacts introduced by the technique. Since forensic scientists would use microbial community typing to compare soil samples, rather than fully characterize their components, artifacts are acceptable so long as they occur predictably and do not impact on the ability to make accurate comparisons.

Of the current DNA-based techniques available, only PCR-based methods have the sensitivity necessary for forensic laboratories. Since most forensic science laboratories have access to automated DNA sequencers, AFLP, A-RISA, and T-RFLP stand out as methods that can accurately and sensitively distinguish between microbial communities. AFLP has the advantage of capturing a wider range of soil community members due to the nonspecific range of its primers, and is also relatively easy to standardize, due to the availability of kits. A-RISA provides a sensitive analysis, combined with the shortest turnaround time of the three methods. This could be an especially important consideration when the techniques find use in a high-throughput laboratory environment. Finally, T-RFLP has the greatest history of use for analysis of soil microbial communities and as such, has gone through extensive scientific, if not forensic, validation. Additionally, the ability of T-RFLP to target specific genes other than 16S rDNA may be important to future forensic science applications.

Although modern soil DNA analysis techniques are certainly applicable and powerful, it is still important to bear in mind the basic physical techniques used to compare soil samples. As noted before, analysis of soil color alone can distinguish a majority of soil samples, and requires little in the way of sample, equipment, time, or expertise. If analysis of soil color and other basic physical tests can exclude a connection between a suspect and a crime scene or victim, it is prudent to conduct these tests before considerable time and expense are devoted to DNA analysis for a case.

10.7 Problems That Remain to be Solved

It is important to recognize that many crucial questions remain to be answered before microbial community typing of soil can be critically evaluated for probative value in forensic applications. Many of the questions that must be answered are also of interest to microbial ecologists and soil biologists, so extensive studies may be anticipated from multiple scientific directions. One very significant problem is establishing the expected variability of microbial soil communities over space and time. For forensic casework, human/plant/animal evidence is straightforward; samples may degrade over time, but individuals will not change STR profiles. While some casework will present mixtures of individuals, the numbers of contributing genotypes are relatively small. Such statements cannot be made of microbial communities, since a DNA profile represents the sum of all individuals in the population. Its composition, and resultant community DNA profile, is expected to change with changing environmental conditions.

Because of the dynamic nature of living components in soil, even storage conditions must be studied and standardized. For example, if a protocol called for samples to be stored in a cool, dry place until extraction, the microbial communities could shift to contain more members that favor growth under just such conditions. In this example, storage conditions themselves exert an unintended selective pressure to standardize the communities within soil samples, resulting in apparent similarity. For these reasons, DNA extraction would need to be conducted as soon as a sample is obtained, or storage conditions developed that captured the community extant at the time of collection.

Actual variation between soil communities in the field is another issue of importance. Extensive research will be required on the effects of spatial, temporal, and climatic variables before any interpretations of the meaning of a "match" or "exclusion" based on soil DNA can be made. T-RFLP profiles can differ over a distance as short as 1 meter between soil and sediment.[27] Information about the changes of microbial community composition as a function of the depth of collection will be required, as will seasonal changes. Given the number of bacteria that form symbiotic relationships within the rhizosphere, it is possible that the microbial community could vary dramatically, even as samples were taken from beneath two separate neighboring plants. Even within one area, the community structure will change over time, on a grand scale as pastureland reverts to forest[32] and in the short term from renovation and industrial development.[27] For the forensic scientist, this emphasizes the critical importance in gathering samples as quickly as possible once a crime scene is discovered. These fluctuations could undermine the effectiveness of a soil profile database, unless it was shown that the areas included in the database had stable microbial communities, or at least communities that varied predictably over the seasons.

10.8 Prospects for Forensic Applications of Soil DNA Typing

Despite these drawbacks, it is clear that DNA analysis of soil samples could, in the future, provide the forensic science community with another robust technique to link suspects to crime scenes or victims. Of primary importance in forensics is to choose and standardize a technique so that analyses can be made reproducible between laboratories. Databases would be required to access the validity of soil DNA typing, to check how samples vary across time and space, and to develop statistical confidence factors and probabilities. This would also help in revealing potential problems with the methods as they occur, such as whether or not certain means of storage impact results.

A future application worth consideration is the use of soil DNA to provide possible investigative leads for a variety of forensic situations. Soil from grave sites can vary significantly in microbial activity from ordinary soil.[35] If genes could be identified that correlated well with the activity of decomposers found in grave soils, it is possible that a grave site could be discovered through T-RFLP analysis. Similarly, there are a variety of chemicals associated with clandestine drug laboratories, such as phenyl-2-propone, that frequently show up as contaminants in soils.[36] By targeting T-RFLP analysis to genes used in the metabolism of such contaminants, forensic scientists could detect traces of these chemicals in the soil and get an idea of where the drug laboratories operate and what kind of drugs are being produced. This approach could also be extended to indirectly detect trace chemicals of interest to forensic scientists, such as propellants used in arson crimes. Although DNA methods may not always be the most straightforward means to uncover the presence of chemicals in the soil, it is possible that the combination of automation, multiplexing, and small sample usage that can be achieved with T-RFLP analysis will mean that an entire battery of T-RFLP tests measuring a number of variables could be conducted rapidly and with little additional effort.

Aside from bacterial soil communities, forensic biologists may be able to extend soil analysis to other organisms as well. Fungal colonies, as well as bacterial colonies, have been studied in comparing soil organisms through laboratory culturing approaches.[5] Recently, A-RISA analysis has been used to study fungal communities in soils using different primers from those used for bacterial analysis.[37] Although fungal differences between soils were found to vary less than differences in the bacterial flora, such analysis could provide another useful variable for comparing some soil types. In addition, nematode worms provide an attractive target for soil DNA typing (D. Fitch, personal communication). Increasing the number of taxa screened in soil theoretically improves the resolving power in a manner analogous to adding increasing numbers of STR loci to human, plant, or animal identification strategies.

Modern genetic analysis provides a method of soil sample comparison to forensic scientists with a number of advantages over other techniques. Not only do these methods have remarkable discriminating power and the ability to work with trace materials samples, but they also use much of the same equipment and expertise that forensic laboratories have in-house for use in human DNA typing. Soil analysis also presents scientists with a technology that potentially adapts well to database technology and that can be adapted to be of use in a variety of crimes. It is clear that once the field standardizes its methodology, soil evidence may take on a new importance in providing the justice system with valuable links between suspects and crime scenes.

References

1. Cox, R.J., Peterson, H.L., Young, J., Cusik, C., and Espinoza, E.O., The forensic analysis of soil organic by FTIR, *Forens. Sci. Int.*, 108, 107–116, 2000.
2. Horswell, J., Cordiner, S.J., Maas, E.W., Martin, T.M., Sutherland, Speir, T.W., Nogales, B., and Osborn, A.M., Forensic comparison of soils by bacterial community DNA profiling, *Forens. Sci.*, 47, 350–353, 2002.
3. Sugita, R. and Marumo, Y., Validity of color examination for forensic soil identification, *Forens. Sci. Int.*, 83, 201–210, 1996.
4. Sugita, R. and Marumo, Y., Screening of soil evidence by a combination of simple techniques: validity of particle size distribution, *Forens. Sci Int.*, 122, 155–158, 2001.
5. Van Dijck, P.J. and Van de Voorde, H., Evaluation of microbial soil identity in forensic science, *Z. Rechtsmed.*, 93, 71–77, 1984.
6. LaMontagne, M.G., Michel, F.C., Jr., Holden, P.A., and Reddy, C.A., Evaluation of extraction and purification methods for obtaining PCR-amplifiable DNA from compost for microbial community analysis, *J. Microbio.l Methods.*, 49, 255–264, 2002.
7. Miller, D.N., Bryant, J.E., Madsen, E.L., and Ghiorse, W.C., Evaluation and optimization of DNA extraction and purification procedures for soil and sediment samples, *Appl. Environ. Microbiol.*, 65, 4715–4724, 1999.
8. Roose-Amsaleg, C., Garnier-Sillam, E., and Harry, M., Extraction and purification of microbial DNA from soil and sediment samples, *Appl. Soil Ecol.*, 18, 47–60, 2001.
9. Tiedje, J.M., Asuming-Brempong, S., Nüsslein, S.K., Marsh, T.L., and Flynn, S.J., Opening the black box of soil microbial diversity, *Appl. Soil Ecol.*, 13, 109–122, 1999.
10. Muyzer, G., DGGE/TGGE: a method for identifying genes from natural ecosystems, *Curr. Opin. Microbiol.*, 2, 317–322, 1999.
11. Lee, D.H., Zo, Y.G., and Kim, S.J., Nonradioactive method to study genetic profiles of natural bacterial communities by PCR-single-strand-conformation polymorphism, *Appl. Environ. Microbiol.*, 62, 3112–3120, 1996.
12. Schwieger, F. and Tebbe, C.C., A new approach to utilize PCR-single-strand-conformation polymorphism for 16S rRNA gene-based microbial community analysis, *Appl. Environ. Microbiol.*, 64, 4870–4876, 1998.

13. Moeseneder, M.M., Arrieta, J.M., Muyzer, G., Winter, C., and Herndl, G.J., Optimization of terminal-restriction fragment length polymorphism analysis for complex marine bacterioplankton communities and comparison with denaturing gradient gel electrophoresis, *Appl. Environ. Microbiol.*, 65, 3518–3525, 1999.

14. Andersen, P.S., Jespersgaard, C., Vuust, J., Christiansen, J., and Larsen, L.A., High-throughput single strand conformation polymorphism mutation detection by automated capillary array electrophoresis: validation of the method, *Hum. Mutat.*, 21, 116–122, 2003.

15. Franklin, R.B., Taylor, D.R., and Mills, A.L., Characterization of microbial communities using randomly amplified polymorphic DNA (RAPD), *J. Microbiol. Methods.*, 35, 225–235, 1999.

16. Sikorski, J., Jahr, H., and Wackernagel, W., The structure of a local population of hytopathogenic *Pseudomonas brassicacearum* from agricultural soil indicates development under purifying selection pressure, *Environ. Microbiol.*, 3, 176–186, 2001.

17. Wikström, P., Andersson, A., and Forsman, M., Biomonitoring complex microbial communities using random amplified polymorphic DNA and principal component analysis, *FEMS Microbiol. Ecol.*, 28, 131–139, 1999.

18. Franklin, R.B. and Mills, A.L., Multi-scale variation in spatial heterogeneity for microbial community structure in an eastern Virginia agricultural field, *FEMS Microbiol. Ecol.*, 44, 335–346, 2003.

19. Franklin, R.B., Garland, J.L., Bolster, C.H., and Mills, A.L., Impact of dilution on microbial community structure and functional potential: comparison of numerical simulations and batch culture experiments, *Appl. Environ. Microbiol.*, 67, 702–712, 2001.

20. Liu, W.T., Marsh, T.L., Cheng, H., and Forney, L.J., Characterization of microbial diversity by determining terminal restriction fragment length polymorphisms of genes encoding 16S rRNA, *Appl. Environ. Microbiol.*, 63, 4516–4522, 1997.

21. Horswell, J., Cordiner, S.J., Maas, E.W., Martin, T.M., Sutherland, K.B.W., Speir, T.W., Nogales, B., and Osborn, A.M., Forensic comparison of soils by bacterial community DNA Profiling, *Forens. Sci.*, 47, 350–353, 2002.

22. Clement, B.G., Kehl, L.E., DeBord, K.L., and Kitts, C.L., Terminal restriction fragment patterns (TRFPs), a rapid, PCR-based method for the comparison of complex bacterial communities, *J. Microbiol. Methods,* 31, 135–142, 1998.

23. Engebretson, J.J. and Moyer, C.L., Fidelity of select restriction endonucleases in determining microbial diversity by terminal-restriction fragment length polymorphism, *Appl. Environ. Microbiol.*, 69, 4823–4829, 2003.

24. Osborn, A.M., Moore, E.R., and Timmis, K.N., An evaluation of terminal-restriction fragment length polymorphism (T-RFLP) analysis for the study of microbial community structure and dynamics, *Environ. Microbiol.*, 2, 39–50, 2000.

25. Egert, M. and Friedrich, M.W., Formation of pseudo-terminal restriction fragments, a PCR-related bias affecting terminal restriction fragment length polymorphism analysis of microbial community structure, *Appl. Environ. Microbiol.*, 69, 2555–2562, 2003.
26. Marsh, T.L., Terminal restriction fragment length polymorphism (T-RFLP): an emerging method for characterizing diversity among homologous populations of amplification products, *Curr. Opin. Microbiol.*, 2, 323–327, 1999.
27. Bruce, K.D., Analysis of *mer* gene subclasses within bacterial communities in soils and sediments resolved by fluorescent-PCR-restriction fragment length polymorphism profiling, *Appl. Envir. Microbiol.*, 63, 4914–4919, 1997.
28. Christensen, J.E., Stencil, J.A., and Reed, K.D., Rapid identification of bacteria from positive blood cultures by terminal restriction fragment length polymorphism profile analysis of the 16S rRNA gene, *J. Clin. Microbiol.*, 41, 3790–3800, 2003.
29. Trotha, R., Reichl, U., Thies, F.L., Sperling, D., Konig, W., and Konig, B., Adaption of a fragment analysis technique to an automated high-throughput multicapillary electrophoresis device for the precise qualitative and quantitative characterization of microbial communities, *Electrophores.*, 23, 1070–1079, 2002.
30. Braker, G., Zhou, J., Wu, L., Devol, A.H., and Tiedje, J.M., Nitrite reductase genes (*nirK* and *nirS*) as functional markers to investigate diversity of denitrifying bacteria in pacific northwest marine sediment communities, *Appl. Environ. Microbiol.*, 66, 2096–2104, 2000.
31. Ranjard, L., Poly, F., and Nazaret, S., Monitoring complex bacterial communities using culture-independent molecular techniques: application to soil environment, *Res. Microbiol.*, 151, 167–177, 2000.
32. Borneman, J. and Triplett, E.W., Molecular microbial diversity in soils from eastern Amazonia: evidence for unusual microorganisms and microbial population shifts associated with deforestation, *Appl. Environ. Microbiol.*, 63, 2647–2653, 1997.
33. Fisher, M.M. and Triplett, E.W., Automated approach for ribosomal intergenic spacer analysis of microbial diversity and its application to freshwater bacterial communities, *Appl. Environ. Microbiol.*, 65, 4630–4636, 1999.
34. Robleto, E.A., Borneman, J., and Triplett, E.W., Effects of bacterial antibiotic production on rhizosphere microbial communities from a culture-independent perspective, *Appl. Environ. Microbiol.*, 64, 5020–5022, 1998.
35. Hopkins, D.W., Wiltshire, P.E.J., Turner, B.D., Microbial characteristics of soils from graves: an investigation at the interface of soil microbiology and forensic science, *Appl. Soil Ecol.*, 14, 283–288, 2000.

36. Janusz, A., Kirkbride, K.P., Scott, T.L., Naidu, R., Perkins, M.V., and Megharaj, M., Microbial degradation of illicit drugs, their precursors, and manufacturing by-products: implications for clandestine drug laboratory investigation and environmental assessment, *Forens. Sci. Int.,* 134, 62–71, 2003.

37. Ranjard, L., Poly, R., Lata, J.C., Mougel, C., Thioulouse, J., and Nazaret, S., Characterization of bacterial and fungal soil communities by automated ribosomal intergenic spacer analysis fingerprints: biological and methodological variability, *Appl. Environ. Microbiol.,* 67, 4479–4487, 2001.

11 DNA Profiling: The Ability to Predict an Image from a DNA Profile

HOLLY LONG, B.S.

Contents

11.1 Introduction

On February 23, 2005, a new service was released that has revolutionized the way DNA may be utilized to solve crimes by increasing the predictive power of a DNA profile. The DNA Witness™ testing services are currently only performed at the DNA Print Genomics Laboratory in Sarasota, Florida. In February of 2005, DNA Print Genomics considered making a kit available to law enforcement agencies so that the testing could be done in their own crime labs, if so desired. At time of publication, this kit has not yet been released to United States Forensic Laboratories. Several other private testing laboratories are also providing single nucleotide polymorphism (SNP) testing to generate a likely facial image to go along with a DNA profile.

The purpose of the DNA Witness SNP test is to provide investigators with the possible ancestry and a predictive image of an individual whose DNA is found at a crime scene. The test kit provides information on the percentage of biogeographical ancestry of four possible genetic groups using a fairly recent discovery in DNA analysis called single nucleotide polymorphisms (SNPs). The four racial groups that have been classified with SNP markers are Sub-Saharan African, Native American, East Asian, and Indo-European. Some of the SNP markers are more prevalent than others (i.e., will give a likely prediction of ancestry) based on the ancestry/ ethnic history of the person tested. The test provides information to limit the possible suspects in which the DNA from a crime scene could have originated.[1]

Although DNA Witness SNP tests have aided in solving a number of crimes that were otherwise going to become cold cases; the test has created much controversy in the forensic and legal fields. There are many ethical concerns that arise when law enforcement agencies begin determining personal information about a suspect based on a DNA profile. Also, the effectiveness of predicting the possible physical characteristics of a suspect is an issue since the tests are not 100% accurate. Law enforcement agencies may waste time and money searching for a suspect based on the results of the testing that may be a false lead. Another issue for debate is the concept of "race." Many scientists are in debate over the idea that separate races in the human population actually exist due to admixture between individuals of different ancestry. This is an important concept for the use of the DNA Witness test, because it theoretically predicts the possible race of an individual and also is the reason why the SNP test in not 100% accurate. In addition, some SNP markers have not yet been identified that can distinguish between all self-identified groups in a population. This type of testing is still in its infancy for human identification; the value for nonhuman identification has yet to be determined by extensive research and screening of relevant populations of different species.

Initially, the purpose of this paper was to determine the usefulness of the test in establishing the possible physical characteristics of a suspect. One issue involved in SNP research is to investigate why individuals from southern India and Europe are grouped together into one racial group when individuals from each population have distinctively different physical features. Therefore, if a suspect is from Europe or from India they could possibly have the same biogeographical makeup by SNP tests. That presents a possible problem for forensic investigators, because individuals from Europe and India have different physical appearances; therefore, it may make it more difficult to locate a suspect. Since it appeared that individuals from Europe and India could not be distinguished based on SNP markers, a research project was proposed to obtain DNA samples from individuals from England and India. The DNA samples would be sent to DNA Print Genomics for the ancestry testing. The results would then have been analyzed to determine if there was some identifying marker to tell the two populations apart.

11.2 The Debate on the Concept of Genetic Race

To begin to understand the usefulness of the products and services provided by a company such as DNA Print Genomics, it is important to understand the concept of "race" and how the human species developed genetic differences over time. There are many misconceptions on the idea of race. The term *race* has not always been used primarily to describe people from different populations. The term was initially used to describe a group of people that shared something in common. In fact, race was used to describe a group or people with similar nations of origin (German race), a similar lifestyle ("a race of women warriors"), or a similar religion.[2]

Europeans began to explore other lands with very different types of populations of residents. Upon the discovery of people with different skin colors and behaviors, Europeans began trying to classify these different people into separate subpopulations of humans. Their attempts led to the development of the "racial worldview." The racial worldview was a set of beliefs that gave people a systematic way to divide humans into different groups. The beliefs were as follows:

- There are separate, distinct, and exclusive populations.
- Phenotypic features, or visible physical differences, mark race, identity, and status, but only "racial essence" need be present to classify an individual.
- Races have distinct behavioral traits.
- Races are unequal and must be ranked.
- Behavioral and physical attributes of each race are fixed, permanent, and unalterable.
- Distinct races should be kept separate and allowed to develop their own institutions.[2]

The scientific classification of humans into distinct groups began in 1735 when Linnaeus developed a classification system for all known animals. Linnaeus called humans *Homo sapiens*, for which he created four subdivisions. The four subspecies were *Homo americanus*, *Homo africanus, Homo europaeus,* and *Homo asiaticus.* Later, a scientist named Blumenbach further classified humans into five types. However, he specified that the groups could not be clearly distinguished because they often blended together. The five groups characterized by Blumenbach were American, Malay, Ethiopian, Mongolian, and Caucasian.[3]

Scientists have concluded through various methods that the modern-day human has evolved from a small founder population of humans in Africa. The oldest known remains of the family Hominidae are thought to have belonged to the *Australopithecus* genus. These remains were found in Africa. It is believed that from this genus evolved the many species of humans. Through millions of years of evolution, *Homo habilis* arose, and until about 1.5 million years ago, *Homo habilis* inhabited sub-Saharan Africa. *Homo erectus* was the next to evolve, and they lived from 2 million to 250,000 years ago. *Homo erectus* slowly migrated to parts of Europe. About

400,000 years ago, it is believed that *Homo sapiens* emerged. *Homo sapiens* were first present in Africa and some parts of Asia, and modern humans are thought to have originated in Africa as well about 100,000 years ago. The modern-day humans eventually migrated to Europe as well. Fully modern humans then evolved about 15,000–30,000 years ago, and they were the inhabitants of many different parts of the world due to migration patterns.[4] As humans began to migrate, changes in the genetic composition of the species occurred. When a species splits into numerous populations that are isolated from one another, they diverge genetically. Changes in the genome may also occur due to mutations, but humans also have to adapt to the climate and food resources.[5]

The oldest humans were thought to have originated in Africa and the Middle East. These humans were thought to have lived 100,000 years ago. Humans then migrated into Asia, where it is believed that the first diversification between non-Africans and Africans occurred at that time.[5] Although humans are thought to have all originated from a small founder population in Africa and then became distinct populations across the globe, there is actually much less genetic variation between populations of humans than among populations of apes. It is believed that this is due to a reduction in the total number of *Homo sapiens*; it is estimated that the total population decreased to about a few thousand to a few hundred individuals. Therefore, the traits in such a small population size could have been established as a unique set of genetic traits and inherited by the next generation.[6]

Such a drastic decrease in the population size is known as the genetic "bottleneck effect." Scientists believe that a major demographic bottleneck occurred as the human populations passed over the landmass that existed where the Bering Strait now exists. When a population size drastically decreases to only a few individuals, the genetic composition may drastically change as well. Any mutations or rare genotypes that were present in those few individuals may become dominant and frequently expressed. The change in genetic composition usually occurs in small populations that are physically isolated from other populations, but it can also occur in populations that are genetically isolated due to not interbreeding into other groups. The effect of a small population size creating a new genetic composition is also called the Founder's Effect and is due to a bottleneck.[5]

The idea that humans evolved from a single population in Africa has been described as the "out of Africa" model or the single-origin theory. This theory of human emergence suggests that the diversity in the anatomy of hominid fossils represents a diversity of species. Molecular geneticists have studied this model to see if the modern *Homo sapiens* originated in Africa based on studies of bone structure and other anatomical characteristics. It is believed that *Homo sapiens* did not appear in Africa before 160 thousand years ago. The results of mitochondrial DNA testing support the "out of Africa" model.[6] Studies were conducted using mitochondrial DNA (mtDNA), because it does not recombine between generations and accumulates changes very rapidly. The accumulation of changes can suggest "recent" evolutionary changes. Samples of mitochondrial DNA were taken from populations all over the world. The results of mtDNA comparisons suggested that modern humans were descendants of a small population that existed about 150 thousand years ago. Mitochondrial DNA studies from Africa show greater genetic variation, which suggests that Africans have been diversified for a longer period of time. Also, samples of mitochondrial DNA from individuals from Europe and Asia represent subsets of African mitochondrial DNA types. The mtDNA similarities suggest that these people originated from populations in Africa.[6]

During the 20th century, scientists continued to search for a method of determining whether distinct human races existed and could be classified. One of the first scientific studies during this time was conducted using blood proteins. It was discovered through research that the majority of American Indians had a blood type of O; however, there were some exceptions. A few Canadian tribes exhibited type A blood. Scientists explained this observation by suggesting that a genetic bottleneck had occurred as the Native Americans crossed the landmass that existed before the Bering Strait existed. The individuals that remained were most likely blood type O. However, another argument was that natural selection had created the common blood type in the Native Americans. Type O individuals were shown to be more resistant to syphilis, and if they were infected, they showed a much faster response to treatment. Also, certain blood types were more prone to other types of diseases as well.[5] With the discovery of DNA techniques for analyzing the genome, genetic studies were conducted on several genes to determine if there were any genetic differences between or within different populations of humans. Studies that were conducted on genes controlling blood group proteins showed that there were some distinct differences between certain human populations. However, it was also determined that no two distinct races have completely different genetic compositions, even for one gene[5], so the search for a race-definitive "marker" continues to this day. This discussion of how best to define a genetic race relates to not only human classification systems but to nonhuman species that have high levels of genetic inbreeding or have experienced genetic bottlenecks as well.

The first DNA-based genetic studies to determine race were conducted on the GC gene, which codes for a blood protein that binds vitamin D and controls vitamin D distribution throughout the body. The two major

Table 1 Distribution of Genes in Different Human Populations

Gene Type	Europe	Sub-Saharan Africa	India	Far East	South America	Australia
GC-1	72%	88%	75%	76%	73%	83%
GC-2	28%	12%	25%	24%	27%	17%
HP-1	38%	57%	17%	23%	60%	27%
FY-O	0.3%	87%	3%	0%	0.2%	0%

forms of the gene are GC-1 and GC-2. Other genes that were studied included the Haptoglobin (HP) gene and the Duffy (FY) gene. HP-1, HP-2, and HP-O are genes that code for the blood protein that is responsible for binding hemoglobin that is released when a red blood cell is destroyed or decays naturally. The FY gene is responsible for the FY-O substance that is found on the surface of red blood cells. The FY substance is a protein that allows the entry of certain parasites into the red blood cells. One of the most common parasites is the malarial parasite. Therefore, certain individuals that do not express the FY gene have a much lower incidence of malaria. Table 1 summarizes the distribution of the GC, HP, and FY genes in different populations. As demonstrated by the table, there is little variation between populations in the GC gene. There is some variation with the HP gene, but the greatest variation is seen with the FY gene.[5]

Based on the results from these studies, scientists were anxious to find other methods that could be utilized to prove that different races existed in the human population. Attempts were made to study the genetic differences that may exist in humans by studying the architecture of certain ancient civilizations. It was believed that if humans that were geographically or genetically isolated had similar architectural styles, they must share some genetic similarities. However, Luca Cavalli-Sforza and colleagues conducted many studies on the relationship between architectural styles and genetics and found no correlation between the two. Therefore, the researchers turned to other characteristics of populations including agricultural developments and language.[5] Although geographical, linguistic, political, and cultural barriers can isolate populations and lead to genetic differences between human populations, Cavalli-Sforza came to the conclusion that a pure race does not exist in the human population. In order for a pure race to be established, there must be 20–30 generations produced by the mating of a father with a daughter or brothers and sisters mating. Even then, there would still be some detectable genetic variation.[5] According to scientific studies conducted in the 20th century, two facts were established:"There is no genetic indicator that can be used to divide populations into races, and geographically distant populations that are the basis of current race classifications vary from one another only in about 6% of their genetic make-up."[2]

However, in 2003, a new genetic technology contradicted the tests that were conducted in prior years. Subsections of the human population occurred as the human species continued to diverge as groups in East Africa migrated to various places around the globe. Humans from different populations around the world are thought to have originated from the same area because there are alleles (i.e., forms of a gene) in the human genome that are similar among all populations. Also, the frequency of alleles in one population is generally similar to all other populations. However, there are certain alleles or genetic markers that are more common in one human population than in others. These markers have been named ancestry informative markers (AIMs). AIMs can be used to predict the possible ancestral origins of a person or a population.[7] Depending on which genetic markers you use for classification, you may be able to detect differences between races within the human population.

It was discovered that single nucleotide changes at certain markers were more common within some human populations than among the populations used in the studies. DNA Print Genomics discovered a set of single nucleotide polymorphisms (SNPs) that could be utilized to potentially predict the ancestry of an individual by indicating the population to which the individual belonged. Researchers at DNA Print Genomics analyzed SNP markers in genes that were responsible for skin pigmentation and xenobiotic metabolism in humans. The researchers chose these genes because it was believed that these genes underwent greater evolutionary selection over time. These genes are involved in ultraviolet radiation protection, dietary tolerances, and possibly physical appearances. Therefore, the environment in which the populations lived most likely influenced the inheritance pattern of these genes. In total, 211 SNP markers in these genes were analyzed. The researchers discovered that the frequencies of alleles in 56 SNP markers showed distinct variation between populations of individuals from European, African, and Asian descent. It is important to mention also that the individuals studied were not genetically related. The SNP analysis found that 15 of the SNPs analyzed were the most informative, but for accuracy purposes in forensic testing, all 56 SNPs had to be analyzed for each sample.[8]

Once these SNPs and AIMs were discovered, the company, DNA Print Genomics, developed certain tests to analyze the AIMs and determine an individual's biogeographical ancestry (BGA). BGA is defined by the company as "the biological or genetic component of race," which means that the BGA is heritable portion of race.[9] The BGA is used to represent a person's ancestry by determining the percentage of ancestry, which a person may possess from one or more of the four major population groups: Native American, European,

East Asian, and African. Individuals belonging to the Native American group historically migrated from Asia to North, South, and Central America. Europeans are classified as those individuals that historically originated in South Asia, the Middle East, and Europe. However, the European category also includes individuals from the Indian subcontinent, which includes Pakistan, India, and Sri Lanka. The East Asian category includes individuals from Southeast Asia, China, Mongolia, Korea, Japan, the Pacific Islands, and the Philippines.[10]

In order to understand the usefulness of the DNA Witness testing kit, one must have a brief understanding of the different options available for the utilization of DNA in the forensic field. There are various methods for examining DNA samples. Some methods are utilized for determining the identity of an individual, while other methods are not individualizing. Some of the most common types of DNA analysis conducted in forensic science are short tandem repeats (STRs), mitochondrial DNA analysis (mtDNA), and single nucleotide polymorphisms (SNPs).

DNA typing or DNA fingerprinting was first discovered by Alec Jeffreys in 1985. Dr. Jeffreys observed that certain portions of DNA were repeated over and over again in the human genome. He also discovered that some of these sequences were highly variable from individual to individual. The repeating sequences became known as variable number of tandem repeats (VNTRs). Dr. Jeffreys developed a typing technique called the restriction fragment length polymorphism (RFLP) method. The RFLP method used restriction enzymes to cleave the DNA in specific areas around the VNTRs. The cleaved portions could then be visualized using gel electrophoresis and southern blot analysis. By comparing the size of the DNA fragments between questioned sample and the known sample, it could be determined whether the questioned sample originated from a certain individual. To give a higher power of discrimination between individuals, multiple loci are analyzed with the RFLP method. Although the RFLP method is very discriminating, it is time consuming, labor intensive, and not easily automated. Therefore, it is harder to analyze multiple loci at one time, which is desired when processing many DNA samples at once. Another disadvantage to the RFLP method is determining mixtures, since the patterns produced by the bands may be difficult to interpret.[11]

Therefore, scientists developed STR markers to analyze DNA samples. STR markers can determine the identity of an individual in a similar fashion as the RFLP method. STRs can be multiplexed or combined for analysis.[12] Utilizing multiplexed STR tests is a huge advantage for a laboratory because one DNA test can analyze multiple STR markers simultaneously. The method is easily automated, highly discriminating, can be utilized for degraded samples, and can decipher sample mixtures.[11]

Mitochondrial DNA analysis is the least discriminating, because it is inherited maternally. Therefore, every individual in the same maternal line will have identical mitochondrial DNA (with the exception of any mutations that may occur). For example, a son will have the same mitochondrial DNA as his mother, whose mitochondrial DNA will be identical to her mother's, and so on. Mitochondrial DNA is useful because it can be extracted from highly degraded sample such as bones, hair, and teeth. The extracted mitochondrial DNA can be compared to individuals in a maternal line to determine if the individuals are related to the same family group. Therefore, mtDNA cannot be used to determine the exact identity of an individual, only that the individuals came from the same maternal line.[11]

11.3 The Use of SNP Markers

Typing by single nucleotide polymorphisms (SNPs) is a very unique form of DNA typing. Within a certain point in the genome, there can be a change in a single nucleotide base that varies between individuals. SNPs are useful in predicting a person's possible race, because SNPs have a much lower mutation rate than STRs, and they tend to become "fixed" in a population. According to scientific calculations, it is estimated that SNPs change on the order of every 10^8 generations. STRs, on the other hand, change every one in a thousand generations. There are certain rare alleles as well that may help to determine ethnicity.[13] Test results include statistical algorithms to calculate a fairly accurate percentage of an individual's ancestry. When the results of the test are reported, the individual receives a copy of his or her genotype, which shows the sequences of DNA that were analyzed along with the location of the markers that were analyzed. Every individual will receive two sequences, because all of the chromosomes are analyzed, and an individual receives one chromosome from his or her mother and one chromosome from his or her father. Therefore, each sequence is bi-allelic, meaning that there are two letters (i.e., possible nucleotide bases) at each marker. The positions that are analyzed vary from individual to individual and from population to population.[7]

Additionally, the results of the ancestry test are reported as a triangle plot. An equilateral triangle is constructed, and each vertex of the triangle represents one of the four major population groups. The triangle plot allows an individual to visually understand his or her maximum likelihood estimate (MLE), which is the estimate that most accurately represents one's ancestral proportions. On the triangle plot, the MLE is represented by a red dot. A perpendicular line is drawn from each vertex of the triangle to the edge of the triangle below it. The perpendicular line will be used as a scale to measure

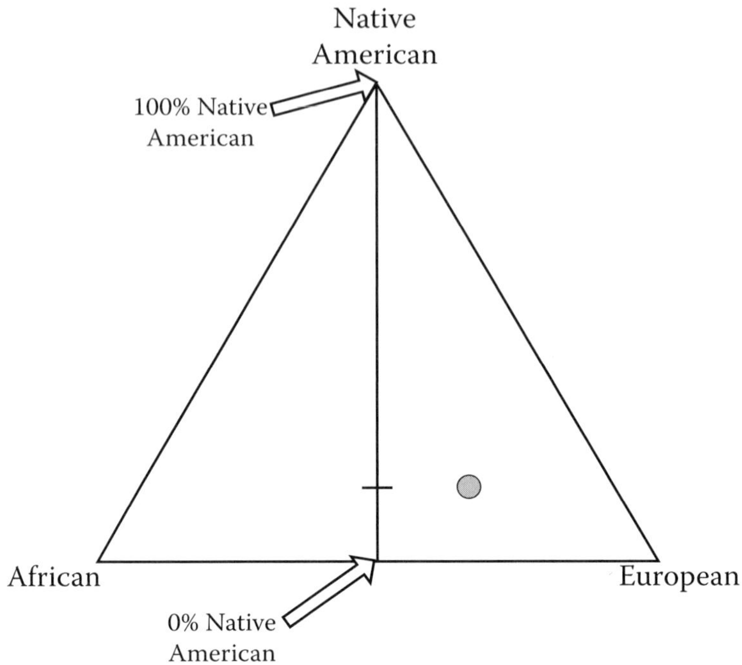

Figure 1 Triangular plot to estimate BGA. (http://www.ancestrybydna.com)

the ancestral proportions. The vertex in which the line is started represents the 100% mark, and the edge of the triangle where the line intersects represents the 0% mark. Therefore, the distance from the MLE dot to the line can be used as a way to measure the ancestral proportions. An example is shown in Figure 1.[7]

The second way in which the results are reported is by bar graphs. The MLE is reported on the bar graph, and confidence intervals are also established and displayed on the bar graph. The proportions are reported with values in the two-fold confidence range as well. Each group is reported separately. Figure 2 demonstrates a bar graph of an individual whose MLE is 55% Native American and 45% European. The graph also shows the other possible proportions of an individual's estimated ancestry, but each of these proportions is at least two times less likely than the MLE. The bars represent the most likely proportions. The lines represent proportions that are least likely. The figure also demonstrates the inability to determine an individual's BGA with 100% accuracy.[7]

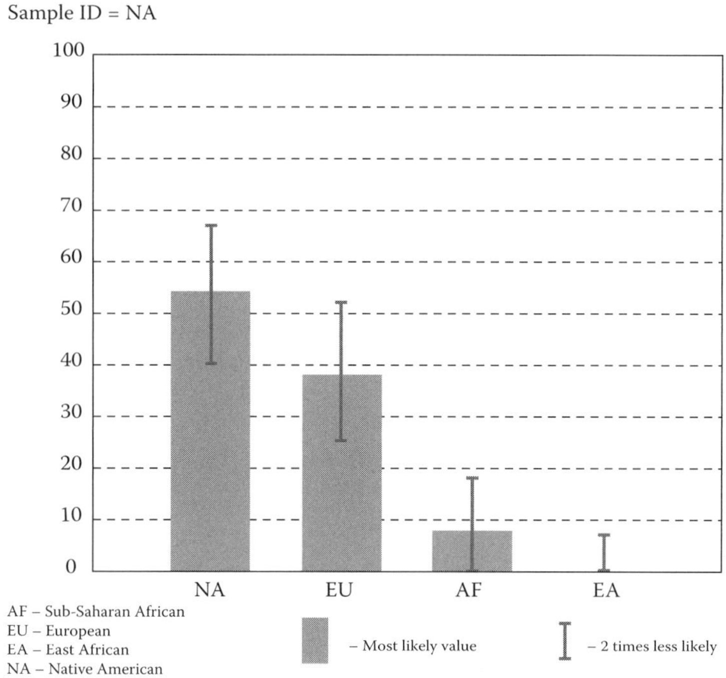

Figure 2 Bar graph representing estimated BGA. (http://www.ancestrybydna.com)

11.4 SNP Testing in the Forensic Field

DNA Print Genomics began providing their DNA Witness test in February of 2003 to law enforcement agencies. The DNA Witness test works on similar principles as the Ancestry by DNA test. The DNA Witness test analyzes certain SNP markers in the DNA found at a crime scene and allows investigators to obtain the percentage of biogeographical ancestry of the suspect, which provides an investigative lead to law enforcement by predicting information about the possible physical features of a suspect.[1] DNA Witness tests have been utilized by customers such as district attorney's offices, sheriff departments, and medical examiner's offices in cities such as New York City, Chicago, and Los Angeles.[14] The test has also been utilized by police in states such as Virginia, Colorado, California, and Missouri, and in the United Kingdom. The DNA Witness test has helped investigators gain leads in over 80 homicides, missing-persons cases, and rapes.[15] According to Matt Thomas, a senior scientist at DNA Print Genomics, the company has conducted 100 DNA Witness tests, with some police agencies using the testing more than once (Matt Thomas, personal communication).

The first case to be aided by the use of the DNA Witness test was one involving a serial rapist named Derrick Todd Lee. DNA at the crime scene of five murders occurring over an 18-month period had been linked to one person but that individual remained unidentified. An eyewitness had told police that she had seen a white male leaving the vicinity of the crime scene in a pickup truck. Police spent valuable time searching for the white male and collected voluntary DNA samples from 1,000 white males in the Baton Rouge area. None of the DNA samples collected from the white males were consistent with the DNA found at the crime scenes. In March of that year, scientists at DNA Print Genomics contacted investigators in the police department. The company had asked the investigators if they could try out their new test that could predict the physical characteristics of a suspect. The results of the experimental test told investigators that the suspect was actually 85% African-American and 15% Native American. The new profile of the suspect led investigators to Derrick Todd Lee, who had a long record of previous criminal activity.[16] The sketch of the individual created from an eyewitness account was very different from the facial features of the defendant, Derrick Todd Lee. The DNA Witness test aided investigators to find the true perpetrator, because the eyewitness identification was incorrect and misled investigators. This case demonstrates how powerful the DNA Witness test may be in criminal investigations. The DNA Print Genomics Company has a database of over 300 photographs of individuals of different races.[16] In addition to the estimated BGA, the company supplies the law enforcement agency with the 100 photos in which the suspect's BGA profile most closely matches (Dr. Coyle, personal communication). In this manner, investigators have an approximation method of determining how the suspect might look by comparing the 100 facial images.

11.5 Ethical Issues

The technology discovered by DNA Print Genomics has great value for criminal investigations. However, there are some issues and ethical considerations that arise from the test results. One issue is that a large proportion of biological forensic evidence taken from a crime scene is a mixture, representing more than one individual's DNA. Secondly, the results of the ancestry test have not yet been ruled admissible in court; BGA test results are used to provide investigative leads and are subsequently confirmed by STR typing.

At a crime scene where biological evidence may be found, many forensic DNA samples may be mixtures. For example, in sexual assault cases, a vaginal swab is obtained from the victim, and if semen is present, the STR profile will be a mixture of the semen depositor and the victim. In STR typing, the analyst can obtain the known profile of the victim and subtract that profile out of the profile obtained from the evidentiary swab. Therefore, it becomes easier to discriminate between the suspect's profile and the victim's profile.[12–14] Obtaining information from a mixed sample with the DNA Witness test is not as straightforward. The company can use the subtraction method if the ancestral proportions of the victim are known. However, the results will still be an estimate of percent BGA because of the inherent limitations of the test.[8] According to Matt Thomas, Senior Scientist at DNA Print Genomics, the company only uses samples from a single source, because they have great difficulty in determining the admixture of a mixed sample. However, if one individual is a minor contributor to the DNA, such as the female DNA in a semen sample collected from a sexual assault case, there are ways to possibly determine the major and minor contributors of the sample (Matt Thomas, personal communication).

In addition to the problems with resolving mixtures, it is also important to consider whether any other typing can be conducted on the evidence as well. If ancestral proportions may be obtained but a STR profile cannot be obtained, the test results may not be very useful because they will not be admissible in court as yet. Although the results of the test may be useful for obtaining valuable information on the physical characteristics of a suspect, law enforcement agencies might be slightly reluctant to use the test if they cannot confirm the results. Degraded and compromised DNA samples have provided SNP results when an STR profile on the same sample was partial or poor. This fact is an issue, because there would be no way to confirm that the sample belonged to the suspect (Matt Thomas , personal communication).

Other ethical considerations involve ultimate usage of tests that can predict an individual's ancestral proportions. Many opponents to the methods used for the SNP-based ancestry test revolve around the issue of determining the race of an individual. Opponents are concerned that ancestry testing may be utilized for discrimination purposes against persons of specific ethnic backgrounds. Also, the risk of other types of discrimination arises as well. SNP markers have been utilized to determine one's susceptibility to some diseases as well. Since the nucleotide sequences of some regions of DNA code for the amino acids that will code for proteins, one nucleotide change at a certain region may change the amino acids, which in turn may alter the production of certain proteins. One example of the involvement of SNPs in disease is the *ApoE* gene, which has been linked to Alzheimer's disease. ApoE can be expressed by three possible alleles, E2, E3, and E4, each one differing by a single nucleotide. Patients with the E4 allele have a much higher risk of developing Alzheimer's disease.[17] Therefore, it is possible that health insurance companies or employers could use the results of an ancestral test to discriminate against an individual by declining insurance coverage or employment.

Another issue with the science behind ancestry testing is that, historically, forensic scientists specifically selected regions of DNA that did not code for proteins. When the DNA Advisory Board (DAB) chose markers to be utilized for forensic DNA typing, they were very cautious about using markers located on genes for fear of profiling in the future. According to Troy Duster, a consultant to the National Human Genome Research Institute, "It would be like going to the NFL and concluding that the DNA marker for sickle cell anemia [associated with African ancestry] makes you a good football player."[18]

However, DNA Print Genomics was not the first group to discover ancestral markers in DNA. According to Ranajit Chakraborty, Director of Cincinnati's Center for Genome Information, scientists had discovered markers that were associated with ancestry when the DNA Advisory Board was determining which markers should be utilized. However, the markers were dismissed as possible options due to concerns even then regarding civil liberties and racial profiling. When DNA Print Genomics announced that they had developed a test to determine ancestry using SNP markers, the scientific community was very surprised. Therefore, many scientists in the forensic science community are cautious of having this technology accepted into the field because it is going against the standards that were established by the DNA Advisory Board in 1997.[18]

One of the main issues that the DNA Witness test will have to overcome is that it is not currently admissible in court. Since the results are not admissible in court, it may lead crime laboratories to doubt the usefulness of the test. However, there is precedence for new DNA typing methods becoming admissible in court. In the future, the DNA Witness test may become admissible in court, thereby providing investigators with another very useful tool to convict suspects. According to Matt Thomas at DNA Print Genomics, several prosecutors have advised the company that the results of the tests will be allowed in court as long as they are used as investigative leads rather than incriminating evidence against an individual (Matt Thomas, personal communication).

The rules for the admissibility of test results and new scientific technologies began in 1923 with *Frye v. United States*. The ruling of this case states that "expert opinion based on a scientific technique is inadmissible unless the technique is generally accepted as reliable in the relevant scientific community."[19] The ruling became known as the Frye test, and it was applied to all evidence. According to the ruling, a new technique or test must exhibit some evidential importance, the technique must be generally accepted, and the theory in which the test is based must be based on generally accepted scientific principles.[20]

Although the Frye test has been subsequently modified, complications arose when trying to apply it test to newly discovered DNA techniques.[20] The methods required for the testing of DNA evidence not only required the actual techniques used to analyze the DNA, but also involved input from other disciplines such as population statistics and genetics in order to interpret the results. Therefore, not only did the techniques undergo the Frye Test, but the actual results did as well.[20] Under the Frye rules, the problem faced when DNA testing became available may also arise if experts seek to allow DNA Witness tests to be admissible in court. The results of the test are based on very new techniques and also on the controversial issue of race. Scientists are still in debate over whether or not race exists. In order for the results of the test to become admissible in court, experts will have to ensure that the results are based on science that is generally accepted by scientists in the field.

The Frye test has historically been a very strict method for allowing certain items or techniques into court as evidence. Therefore, in 1975, the Federal Rules of Evidence (FRE) were devised to give courts some flexibility in determining admissibility. The two major rules that address the admissibility of evidence in court were FRE 702 and FRE 703. FRE 702 states that an expert in a specific field can act as a witness to testify about the results of a certain scientific test or technique if, by explaining the test or technique, the judge and/or jury will better understand the technique as it applies to the evidence. FRE 703 states that the facts or data that lead to an expert opinion do not have to be admissible in court as long as the method on which the facts or data were obtained was generally accepted by other scientists.

The FRE allow judges to consider many other issues regarding the evidence other than the general acceptance standards. Although the FRE were more lenient than the Frye test, a battle arose among courts because some jurisdictions were using the Frye test, while others were utilizing the FRE. Therefore, in 1993, the Daubert ruling determined that the Federal Rules of Evidence superseded the Frye test. After the Daubert ruling, several new principles were set forth regarding the admissibility of new scientific techniques or tests. The Daubert ruling declared, "the trial judge is responsible for the task of ensuring an expert's testimony both rests on a reliable foundation and is relevant to the task at hand."[19] The Daubert ruling differed from the Frye test, because it focused more on the method than the test results.[20] According to the Daubert ruling, a new method must meet the following requirements to be considered admissible:

- The theory or test must have been or have the potential to be tested.
- The theory or technique must have been subjected to peer review and publication.
- The error rate must be known.
- There must be standards for controlling the operation of the method.
- The new method must have attracted widespread acceptance in the scientific community.[20]

Utilizing the FRE standard for the admissibility of evidence, it may be difficult for the results of the DNA Witness test to be admissible in court.

This raises a problem for investigators that may consider using the test. Investigators may be reluctant to use a test that is fairly expensive if the results cannot be used in court. One of the issues regarding the admissibility of DNA evidence was stated in a report in 1992 by the National Research Council. There had been a debate regarding the population match frequencies of the new DNA technology, and courts were reluctant to accept the results of the tests. However, in 1996, the debate was settled when the National Academy of Science press published a report that supported the methods in which the population match frequencies of the new DNA technology were obtained. This report allowed the DNA testing results to be admissible in court because the statistical interpretation of the results was deemed adequate.[19] A similar evaluation and report may be necessary for SNP-based forensic tests for human and nonhuman evidence.

The DNA Witness test may experience similar admissibility issues because the results are statistical proportions as well. Also, the results of the test are not 100% accurate, because it is only an estimate of one's ancestral proportions. There is no way to actually prove if the estimates are accurate. In order to prove this, researchers would have to know every single ancestor of an individual and their genetic composition, which is not a practical experiment to perform. Despite the fact that some scientific hurdles exist to evaluate the full usefulness of the DNA Witness test, there has been enough evidence presented in the form of casework that supports that the technique should be adopted by law enforcement agencies that are concerned with locating a suspect when little or no other information is available. As with any new scientific test, the DNA Witness test has many advantages and some disadvantages, but overall, the DNA Witness test is a very powerful tool.

Some of the major disadvantages of the test kit involve the fact that the results are not 100% accurate. Since the test results only provide investigators with the possible race of a suspect, the investigators could possibly be misled. Another disadvantage involves the true genetic makeup of the individual. Many African-Americans tend to have a large amount of admixture. For example, they usually show at least 20% European ancestry as well. Caribbean Hispanics show a large percentage of European, Native American, and African admixtures. Since individuals from these populations tend to be greatly admixed, it may be hard to match the suspect's DNA in a database. Also, since the percentages are so admixed, it may be difficult to conclude exactly what race the individual may be in some cases. Another disadvantage is the fact that admixtures in Europeans are much harder to interpret. For example, individuals from Russia, Scandinavia, and Eastern Europe tend to have low levels of East Asian admixtures. Also, individuals of Greek, Italian, and Middle Eastern descent usually show a small mixture of American Indian descent even if there is no record of an admixture event in their history.[10]

However, there are many advantages to the test, which make it very useful to law enforcement agencies and forensic scientists. The DNA Witness test is the only test that is capable of predicting any information about a suspect. The test is also greater than 80% accurate in its predictive capabilities. DNA Print Genomics conducted several blind tests on samples submitted from police departments. One test processed 16 samples from members of the police department, and the results were accurate in all 16 samples. Another test was conducted in a separate police department, in which 20 samples were collected and tested. Again, the company was able to correctly predict the population group to which the individual belonged.[14] Another advantage is that this test may help to speed up the process of retesting crime scene samples of individuals who have been incarcerated. Individuals that feel they were wrongfully convicted may utilize this technique to show that they could not have been the suspect and subsequently obtain permission for the crime scene evidence to be retested to prove that they are innocent. For example, if a test comes back saying that the DNA belongs to an individual of African-American ancestry and the individual that is incarcerated is of Native American descent, the incarcerated individual may have a better argument for having STR testing conducted.[14]

The DNA Witness tests are also advantageous to investigators, because often there are no eyewitnesses to a crime to provide any information on the suspect. The DNA Witness test may act as a silent witness by providing investigators with some information on the suspect's physical appearance. The information provided by this test might be extremely valuable, because it can greatly narrow the list of possible suspects, redirect an investigation, and save investigators valuable time. The DNA Witness test requires only a small amount of sample. Therefore, crime laboratories do not have to worry about using their entire sample from a crime scene. The technique also has the ability to be highly automated. The company has been able to obtain 150,000 genotypes in one 8-hour shift. The fact that the test can be highly automated is advantageous to a crime laboratory, because it can save the laboratory a great amount of time and money since there are often many samples that need to be processed. In terms of cost, the test is relatively inexpensive. One test is approximately $1,000, comparable to other types of DNA tests that range from $600 to $1,200 per sample. Another advantage is that the company has modified the DNA Witness test to contain an assay for determining the eye color of a suspect.[16] The DNA Witness 3.0 test has the ability to determine eye color, and researchers hope to add more useful physical traits at a later date.[14] Ultimately, this technology could be used for nonhuman evidence as well.

Although there is some controversy surrounding the DNA Witness test, it can provide law enforcement agencies with very valuable information. It has been proven from the Derrick Todd Lee case that eyewitness identifications can be incorrect and misleading. The test provides investigators with information about the suspect that relies on scientific results rather than the memory of a witness. Also, investigators should seriously consider using the test in the absence of any leads. Some cases have no leads, and if this test can provide any information at all, investigators at least have some clues about the person who committed the crime. The admissibility issue is important, but in reality, the results of the test may not even need to be used in court. The use of the results of the DNA Witness test is similar to the use of serial-killer profiling or behavioral profiling methods. In fact, polygraph testing is not admissible in court, but the use of a polygraph has been accepted in the scientific community as a method of obtaining information. If the DNA Witness test is utilized as any other presumptive test, the results will not need to be court admissible. The DNA sample that was processed using DNA Witness is always confirmed using traditional STR testing, so the results of the STR tests can be utilized in court as confirmation of identity. Although a new forensic DNA technology, the DNA Witness test has been demonstrated to be a very effective and useful investigative tool for human biological evidence and should be considered a legitimate option for investigators seeking investigative leads.

References

1. DNA Print Genomics Press Release. DNA Print Genomics introduces new product for law enforcement. http://www.dnaprint.com.
2. Race, in *Encyclopedia Britannica,* 15th ed., Vol. 9, Encyclopedia Inc., Chicago, 2005, p. 876.
3. Race. Human evolution: Scientific classifications of race, in *The New Encyclopedia Britannica*, 15th ed., Vol. 18, Encyclopedia Inc., Chicago, 2005, p. 847.
4. Human evolution, in *The New Encyclopedia Britannica*, 15th ed., Vol. 6., Encyclopedia Inc., Chicago, 2005, pp. 135–136.
5. Cavalli-Sforza, L.L. and Cavalli-Sforza, F., *The Great Human Diasporas: The History of Diversity and Evolution*, Addison-Wesley Publishing Company, 1995.
6. Human evolution: Body structure, in *The New Encyclopedia Britannica,* 15th ed., Vol. 18, Encyclopedia Inc., Chicago, 2005, p. 842.
7. AncestrybyDNA: User Manual. http://www.ancestrybydna.com/welcom/productsandservices/ancestry.
8. Frudakis, T., Venkateswarlu, K., Thomas, M.J., Gaskin, Z., Ginjupalli, S., and Gunturi, S., A classifier of SNP based inference of ancestry, *J. Forens. Sci.,* 48, 771–778.
9. AncestrybyDNA- FAQs. http://www.ancestrybydna.com/welcome/faq/.

10. AncestrybyDNA- home. http://www.ancestrybydna.com.
11. Butler, J.M., *Forensic DNA Typing: Biology, Technology, and Genetics of STR Markers,* 2nd ed., Elsevier Academic Press, Boston, 2005, pp. 2–5.
12. Dorak, M.T., Glossary of Real-Time PCR Terms. http://dorakmt.tripod.com/genetics/glosrt.html.
13. Butler, J.M., *Forensic DNA Typing: Biology, Technology, and Genetics of STR Markers,* 2nd ed., Elsevier Academic Press, Boston, 2005, 156–189.
14. DNAPrint Genomics – Forensics. http://www.dnaprint.com/2003/services/forensics.html.
15. USAToday, DNA test offer clues to suspect's race. http://www.usatoday.com/news/nation/2005-08-16-dna_x.htm.
16. Genome News Network. Genome Test Nets Suspected Serial Killer. http://www.genomenewsnetwork.org/articles/06_03/serial.shtml.
17. SNP Fact Sheet. http://www.ornl.gov/sci/techsources/Human_Genome/faq/snps.shtml#risks.
18. DNA Witness: Determine the Suspect's Ra ce. http://boards.aetv.com/ thread.jspa?threadID=300001469&tstart=0&mod=1147483398987.
19. Hramon, R., Admissibility standards for scientific evidence, in *Microbial Forensics,* Breeze, R.G., Budlowe, B., Schutzer, S.E., Eds., Elsevier Academic Press, Boston, 2005, pp. 382–386.
20. Palmbach, T. and Shutler, G., Legal considerations for acceptance of new forensic methods in court, in *Forensic Botany: Principles and Applications to Criminal Casework,* Coyle, H.M., Ed., CRC Press, New York, 197–216.

APPENDIX: A Brief History of Forensic Serology and DNA

PETER BILOUS, PH.D.

Contents

Forensic Science and Individualization of Physical Evidence

The primary goal of forensic scientists is to identify, and if possible, individualize forensic evidence to a specific source or to a common source. The former can be achieved in situations of a direct physical match and with certain classes of physical evidence such as pattern evidence and biological evidence. Examples of a direct physical match (also called a *jigsaw* fit) include glass fragments from a breaking-and-entering case, and paint chips from a hit-and-run case. Pattern evidence includes fingerprints, footwear impressions, tire impressions, and impression evidence associated with cases involving tool marks and fired rounds of ammunition.

As a result of the advances in the biological and medical sciences in the 1970s and 1980s, biological evidence has joined the ranks of physical evidence that can be individualized, thanks to the introduction of forensic DNA typing analysis in the mid-1980s. Biological evidence is not only associated with violent crimes of homicide and sexual assaults, but can be found on cigarette butts, food items, stamps, ski masks, steering wheels, windshields, and many other items associated with robberies, threatening letters, theft cases and so forth.

Obtaining a match with these categories of forensic evidence is not a guarantee. The quality of the forensic evidence must be sufficiently high to allow for a thorough examination and comparison to be made with a known or reference sample. The discovery of variability in blood composition was a result of studies conducted to understand the difficulties associated with performing blood transfusions in the 1800s. Karl Landsteiner, an Austrian physician working at the Vienna Pathological Institute, discovered the polymorphic nature of red blood cell antigens in 1901. Landsteiner designated the three major types of antigens as A, B, and O. This was a significant finding, particularly for those individuals requiring blood transfusions. In 1902, Max Richter and Karl Landsteiner proposed that ABO typing could be used in forensic situations.

Leone Lattes, a scientist at the University of Turin, Italy, developed methods to examine proteins in dried bloodstains. These findings were particularly significant to the forensic sciences, as blood evidence is, more often than not, found in a dried state at crime scenes. In the 1920s, it was shown that about 80% of the human population secretes the ABO antigens into other body fluids. These individuals are known as *secretors* and are of value in cases involving fluids other than blood, such as seminal fluid in sexual assault cases. Forensic serology blossomed in the 1960s with the discovery of numerous protein markers and isoenzymes in blood and other body fluids. Numerous polymorphic proteins systems, stable in dried blood, were being examined in an attempt to individualize biological evidence and source it to a particular individual.

The forensic analysis of polymorphic proteins in blood and semen stains yielded random match probabilities (RMPs) generally within the range of 1 in 100 to 1 in 1,000. Although moderately discriminating, the tests for polymorphic proteins were primarily exclusionary in nature. By examining protein polymorphisms, forensic scientists were indirectly examining DNA differences between individuals, as all proteins are coded by DNA. The development of DNA typing procedures in the mid-1980s was the beginning of the end for the classical serological approach of individualizing body fluid stains.

DNA is the genetic material found in all living organisms. DNA contains coding regions (genes) and noncoding regions. Each gene codes for a specific protein. Proteins have a variety of roles within the human body: as hormones (e.g., insulin), as structural proteins (e.g., collagen), and as catalyzers of chemical reactions (e.g., enzymes). DNA is located in all nucleated cells of the human body, but not present in red blood cells, as these cells are devoid of nuclear material. DNA is the same from cell to cell, with the exception of the sperm and the egg, each of which contains only half of the genetic information of the other human cells.

DNA testing has several advantages over the classical protein approach:

Any biological material can be used for testing (blood, semen, saliva, tissue).

DNA is stable if kept in a dry state and away from the harmful UV rays of direct sunlight.

DNA tests are sensitive, requiring only trace amounts of biological material (e.g., saliva on cigarette butts, cells from a latent fingerprint, a single human hair root, pin-head sized bloodstains).

Forensic DNA tests are based on the undisputed premise that everyone's DNA is unique. Only maternal twins (monozygotic) would have identical DNA. An early challenge for forensic scientists was to develop a testing procedure that yielded a DNA profile that with as much certainty as possible represents our genetic uniqueness. Well-characterized highly polymorphic loci were chosen by the major law enforcement forensic laboratories in North America for the first forensic DNA testing procedures. These tests examined polymorphic tandem repeat loci known as minisatellite DNA. The testing procedure was termed RFLP-VNTR DNA typing.

Medical Genetics and Polymorphic DNA Markers

Developments in molecular biology and the discovery of polymorphic genetic markers laid the foundation for their subsequent application in forensic DNA typing analysis. Restriction enzymes cut or cleave DNA at specific recognition sequences, which are randomly located throughout the human genome. The set of DNA fragments produced by such a digestion is unique to each individual due to the presence of variable number of tandem repeat (VNTR) loci scattered throughout the genome.

VNTR loci were first used in medical genetics to study disease genes. The analytical approach used was known as restriction fragment length polymorphism (RFLP) analysis of VNTR loci (RFLP-VNTR analysis). If a polymorphic marker is located near a disease locus, the gene can be traced through a family study allowing for possible location and cloning of the disease gene for further study. The examination of polymorphic VNTR loci provided a high degree of discrimination among individuals, and would be used not only for forensic analysis, but also for paternity testing, and as an important medical diagnostics tool (e.g., bone marrow transplantations).

The Beginning of Forensic DNA

Early DNA identification tests were based on procedures borrowed from the scientific fields of molecular biology and medical genetics. The RFLP-VNTR DNA typing procedures examine inherited repetitive DNA regions of the human genome, known as mini-satellite repetitive DNA loci. The polymorphic nature of these loci arises from the variable number of tandem (core) repeats that constitute the alleles at these loci.

The first DNA typing procedure, developed by Alec Jeffreys, used DNA probes that simultaneously detected a multitude of similar repetitive DNA loci across all of the chromosomes (the multi-locus approach). The DNA profile obtained from this approach was a set of DNA bands resembling the bar code patterns found on store products. In North America, a single-locus DNA typing approach was chosen, testing only one locus at a time. Stringent test conditions were chosen to detect alleles at any one locus of a set. The resulting DNA pattern at a given locus was simple one- or two-DNA bands, representing the genotype of the individual at that locus. Thus, a DNA profile resulted from a series of individual tests, each test yielding a DNA allele pattern at one particular mini-satellite locus. Generally from four to six mini-satellite loci were examined in RFLP-VNTR DNA typing.

The single-locus approach had several advantages over the multi-locus approach. Less sample DNA was needed to perform the tests, degraded DNA samples were less problematic, and the DNA profiles were easier to interpret. However, in paternity cases, the multi-locus approach was faster and more discriminating, as many more loci are examined in one test. The forensic RFLP-VNTR DNA typing procedure involves a series of steps. The first step is the chemical extraction of DNA from biological crime scene samples (blood, tissue, saliva). A standard protease digestion of cellular material in the presence of detergents was used. For semen stains from sexual assault cases, DNA from both the victim and the semen donor are expected. Both the FBI and the FSS developed a modified cell lysis procedure that yielded one fraction enriched with the victim's DNA (from epithelial cells) and one enriched in perpetrators DNA (from spermatozoa).

The second step is to determine the quantity and quality of the DNA isolated. For this procedure, a small portion of the isolated DNA is first analyzed by gel electrophoresis, a separation technique, separating the large DNA fragments from any small ones that may be present due to degradation of the sample DNA. The separated DNA is visualized by staining with a fluorescent DNA-intercalating dye, ethidium bromide. A series of DNA standards (known quantities of high quality DNA) are analyzed simultaneously, providing a means to compare and estimate the quantity of DNA in the forensic samples. For the third step, an appropriate quantity of DNA (generally 500 ng) was then digested (*cut*) with a restriction enzyme to produce a set of DNA fragments. (*Hae*III was chosen by most NA forensic laboratories due to its robustness.) The mini-satellite alleles being examined were not cut internally by this restriction enzyme, except in rare situations in which mutations resulted in the creation of an internal restriction site. As a quality control step, the extent of digestion was evaluated using gel electrophoresis, and the digestion step repeated for any poorly cut DNA samples.

The fourth step was to sort the *Hae*III fragments according to size by gel electrophoresis. The resulting array of DNA fragments (small to large) were then transferred to a nylon membrane using a blotting procedure developed in 1975 by Ed Southern. Blotting was done in the presence of NaOH to denature the DNA into its single-stranded form (ssDNA). The ssDNA was immobilized onto the surface of the membrane during the transfer procedure. The immobilized ssDNA could now be tested using a radiolabeled DNA probe specific for one of the mini-satellite loci. Stringent DNA hybridization conditions were chosen to ensure detection of alleles at a specific locus.

The bound radiolabeled DNA probe was visualized using autoradiographic procedures. The resulting autorad showed either one DNA band (a homozygous genotype) or two DNA bands (a heterozygous genotype). The DNA band pattern obtained was only moderately discriminating. Thus, a series of hybridizations were conducted, each evaluating one of several mini-satellites. DNA probes bound to the membrane had to be completely removed by a "stripping" procedure, prior to hybridization with the next DNA probe. Although each round of DNA hybridization resulted in the loss of some of the bound DNA, numerous tests could be performed before weak signal became a problem during autoradiography.

RFLP analysis required a series of individual tests to be performed, thus inherently a slow process, often requiring several months to complete. Shortly after the development of the RFLP DNA typing procedure, PCR-based DNA typing procedures were developed, with the promise of improved turnaround times for cases requiring DNA typing analysis.

Single-Locus VNTR DNA Probes

Approximately 30% of the human genome consists of repetitive DNA sequences, with two main classes, interspersed and tandem. Moderately repetitive interspersed DNA elements consist of mobile genetic elements such as LINES and SINES. LINES are long interspersed nucleotide elements, greater than 500 bp in length. Examples include L1 elements (6 kb full length, with approximately 50,000 in the human genome). SINES are short interspersed nucleotide elements, less than 500 bp in length. Examples include the Alu sequences (290 bp consensus sequence, approximately 1 million copies in the human genome). Tandem repeat DNA sequences include the classical satellite sequences, Alpha I, II, III, IV (classified based on their sedimentation characteristics during separation on CsCl gradients), and the mini- and micro-satellite tandem repeat loci that are dispersed throughout the chromosomes.

The centromeric region of each chromosome contains a tandem array of alpha satellite repeats that vary in size from 0.2 to 10 Mb. Two alpha satellite loci of importance in the RFLP-VNTR DNA typing analysis procedure were DYZ1 and D7Z2. DYZ1 is an alpha satellite locus located on the Y chromosome, consisting of several thousand copies of a 3564 bp *Hae*III tandem repeat. DYZI was used to determine the gender origin of a sample in the RFLP-VNTR DNA typing procedure. The detection of a 3564 bp DNA band in an *Hae*III digest of a forensic sample indicated the presence of Y-chromosomal DNA, and therefore the presence of male DNA in the sample. D7Z2 is an alpha satellite locus located on chromosome 7, consisting of several hundred copies of a 2731 bp *Hae*III tandem repeat. D7Z2 was used as an internal control marker for the RFLP-VNTR DNA typing procedure, as this was a nonpolymorphic locus. The detection of a 2731 bp DNA band in an *Hae*III digest of a forensic sample indicated that proper restriction by the *Hae*III enzyme had occurred, and that the sample DNA had migrated properly during electrophoresis (the absence of any "band-shifting" phenomenon).

D17Z1 is a locus located on chromosome 17 that was used to quantify the amount of human DNA (i.e., primate) isolated from a forensic sample. Human DNA was quantified using a DNA hybridization procedure after binding a small portion of the sample DNA to a membrane surface. DNA probes specific for D17Z1 were labeled with either radioisotopes, such as ^{32}P for autoradiographic detection, or biotin for color or light-based detection methods using a peroxidase assay. The polymorphic nature of alleles at these loci is due to differences in the number of core repeat units that make up the alleles. Hundreds of different alleles were possible. The RFLP-VNTR DNA typing procedure simply determined the lengths of the alleles at these mini-satellite loci.

Microsatellites are also known as short tandem repeats (STRs), polymorphic DNA loci with core repeats consisting of 2–7 bp. STR markers are scattered throughout the genome and occur on average once every 10 Kbp. The short length of the STR alleles (100–400 bp) makes them amenable to the PCR amplification procedures. STRs are less variable than their mini-satellite counterparts, thus requiring the analysis of a greater number to achieve the same discriminating power as obtainable with the RFLP-VNTR DNA typing analysis of mini-satellite loci.

PCR-Based DNA Typing Systems

The ASO (allele-specific oligonucleotide) DNA typing procedure was a reverse dot-blot hybridization assay. A forensic DNA sample was first amplified using the PCR (polymerase chain reaction) procedure. Following amplification, a test portion of the forensic sample was denatured (made single stranded) and incubated in solution with the nylon membrane for hybridization to occur. Nylon membranes were commercially prepared with oligonucleotides fixed to the surface and arranged in a series of dots, each dot having oligonucleotides that recognized (were complementary to) the DNA of a particular allele at the loci being tested. Post-hybridization development (colorimetric or autoradiographic) produced a series of dots, each dot representing an allele in the sample. The procedure was simple and quick — no restriction endonuclease digestion and no gel electrophoresis.

The human leukocyte antigen gene complex is located on chromosome 6 and became the basis for HLA-DQA1 DNA Typing. Of interest to forensic scientists was the class II alpha chain gene known as DQA1, a hypervariable region of 242 bp. Seven common alleles have been detected (1.1, 1.2, 1.3, 2, 3, 4.1, 4.2/4.3), defining 28 genotypes. A commercial kit first manufactured by Cetus Corporation (called AmpliType) used the reverse dot-blot format with a colorimetric detection system to detect the subtle differences in the nucleotide sequence of the HLA DQA1 alleles. The advantages of the HLA DQA1 kit for forensic DNA analysis were its simplicity and speed. However, there were several problems associated with this procedure:

1. Interpretation was visual and subjective.
2. Mixed samples were difficult to interpret.
3. Degraded samples were difficult to interpret.
4. Power of discrimination was low.

The Polymarker kit was introduced in 1994 to complement the existing HLA-DQA1 kit, and to add additional loci to improve on the discriminating power of the tests. This was the first PCR-based multiplex system, analyzing several polymorphic loci simultaneously. As with the original HLA-DQA1 kit, the test was simple to perform and rapid. Similar problems still existed:

1. Mixtures were difficult to interpret.
2. Degraded DNA reduces signal intensities.
3. Dot intensities were imbalanced within a locus.
4. There was allelic drop-out with the LDLR locus.

This kit was expensive and still had a low discriminating power.

D1S80 is a polymorphic mini-satellite locus, characterized by a 16-bp core repeat. Approximately 32 alleles have been identified, ranging in size from 350 to 1000 bp. Sample DNA was amplified by the PCR procedure, and the amplicons (discrete alleles) were separated on polyacrylamide gels and visualized by silver staining. The length of the amplicons was directly proportional to the number of core repeats present in the alleles.

The forensic community currently uses tetra- and pentanucleotide repeat STR loci for human identification work. Di- and trinucleotide repeat STRs have very high stutter percentages (> 30%), which complicates the interpretation of mixtures and therefore, were not further developed for forensic casework. The PCR-STR DNA typing procedure consists of the following steps:

1. extraction and quantification of DNA
2. PCR-STR DNA amplification
3. DNA profile determination
4. various fluorescent-based detection systems used to analyze DNA fragments (e.g., Hitachi FMBIO II gel scanner; ABI 377-PAG; ABI 310-CE)
5. data interpretation

Fluorescence-based detection systems are widely used because they provide a rapid multi-color analysis. Fluorescent dyes are incorporated into PCR primers in order to tag the amplicons. Amplicons are detected using laser light to excite the various fluorescent dye molecules, which then emit visible light of different wavelengths (e.g., blue, green, yellow, and red emission spectra). In PCR-STR DNA typing analysis, only one primer of each STR set is labeled with a fluorescent dye. Different dyes are used to distinguish between the different loci that are simultaneously amplified by the PCR process, a process called multiplexing. Light emission from the various dyes is detected by photosensitive devices (PMTs or CCDs) that convert the energy of the photons to electrical signals. The light intensity is typically reported in arbitrary units known as RFUs (relative fluorescence units).

PCR-STR amplifies alleles that vary in length; these alleles are small (100–400 bp) and are separated electrophoretically on polyacrylamide gels (PAGEs) or by using polymers within a capillary tube (CE). Internal lane standards (DNA fragments of known size) are run in every sample lane. Allelic ladders are run on every gel to provide a reference for size and allele designation and represent a prepared mix of common alleles found in the human population at the STR loci being tested.

Gender tests: amelogenin is a gene on the X and Y chromosomes, that codes for a protein associated with tooth enamel maturation. The locus is used for sex typing purposes in forensic DNA typing kits. The gene on the X chromosome has a 6-bp deletion. With male DNA, PCR amplification results in two DNA bands, one 106-bp allele (from the X chromosome) and a 112-bp allele if the Y chromosome.

The Development of DNA Quality Assurance Guidelines

From TWGDAM to SWGDAM: The FBI hosted the first meeting of the Technical Working Group on DNA Analysis Methods (TWGDAM) in 1988. The first meeting consisted of 31 scientists from the United States and Canada, including participants from publicly funded and private forensic laboratories. The purpose of the meeting was to discuss methods, share protocols, and establish guidelines. In 1989, TWGDAM issued "Guidelines for a Qualtiy

Assurance Program for DNA Restriction Fragment Length Polymorphism Analysis."

This publication was followed in 1990 by "Guidelines for a Proficiency Testing Program for DNA Restriction Length Polymorphism Analysis." Several other guidelines were issued in the subsequent years. TWGDAM changed its name to Scientific Working Group on DNA Analysis Methods (SWGDAM) in 1999, and has continued to issue important guidelines on new DNA technologies.

The Office of Technology Assessment (OTA) is an analytical arm of the U.S. Congress. Its role is to assist legislators in making new laws and regulations. In 1990, the OTA issued a report entitled "Genetic Witness: Forensic Uses of DNA Tests." This was the first major review of forensic DNA typing analyses. The major conclusions of the OTA report were:

1. The scientific basis for DNA typing is sound.
2. DNA tests are sensitive and accurate.
3. When properly performed, DNA tests are reliable.
4. There is a need to establish both technical and procedural standards.

The National Research Council (NRC) is a branch of the National Academy of Sciences, which was organized in 1916 as an advisory group for the federal government. In 1992, the NRC issued a report entitled "DNA Technology in Forensic Science." The NRC report affirmed the soundness of the DNA typing methods, but called for the establishment of Quality Assurance standards. The report also provided recommendations on technical, statistical interpretation, laboratory standards, databanks and privacy, legal considerations, and societal and ethical issues.

The *ceiling principle* was recommended for statistical calculations. The methods used to estimate the random match probabilities (RMPs) of matching DNA profiles came under scrutiny in the trial of *New York v Castro*. The multiplication rule used to calculate profile frequencies assumes that alleles within a locus and between loci are inherited in an independent fashion. Daniel Hartl of Harvard University indicated that since individuals within certain population groups tend to marry within their group, this created population subdivisions. Thus, allele frequencies within these population groups differed from the random population samplings used to create the population databases used by the forensic laboratories to calculate the RMP. The ceiling principle was controversial due to its highly conservative approach and the lack of a scientific basis in its formulation. This controversy resulted in the almost immediate establishment of a second NRC committee on DNA methods, this time chaired by Dr. James Crow, a world-renowned population geneticist from the University of Wisconsin.

In 1996, the second DNA committee issued its report entitled "The Evaluation of Forensic DNA Evidence." The NRC's second report concluded that the calculations of match probability using the ceiling principle were unnecessary and inadvisable. The technology for DNA profiling and the methods used to estimate frequencies and related statistics were reliable and valid. Furthermore, laboratories should adhere to high-quality standards and make every effort to be accredited for DNA work. The 1996 NRC report called for the establishment of a DNA Advisory Board (DAB) to assist with policy decisions. In 1994, the U.S. Congress passed the DNA Identification Act. A DAB was established by the Director of the FBI, consisting of 13 members. Important guidelines on Quality Assurance were issued in 1998 and 1999. Although originally mandated for just five years, the DAB mandate was extended for a sixth year.

As nonhuman DNA evidence is considered and additional DNA-based methods are developed and applied to forensic casework, the forensic community and regulatory boards will need to assess the quality of the scientific methods and species population databases. DNA is a fast-moving field of forensics and as the technology changes, so will the applications and guidelines to keep pace with the science. The future for expanding and adapting human identification methods to nonhuman evidentiary samples from plants, animals, and insects is bright.

Index

A

B

E

M

N

O

P

Q

R

S

X

Y